Full-Stack Web Development with TypeScript 5

Craft modern full-stack projects with Bun, PostgreSQL, Svelte, TypeScript, and OpenAI

Mykyta Chernenko

Full-Stack Web Development with TypeScript 5

Copyright © 2024 Packt Publishing

Group Product Manager: Kaustubh Manglurkar

Publishing Product Manager: Bhavya Rao

Book Project Manager: Sonam Pandey

Senior Editor: Rashi Dubey

Technical Editor: K Bimala Singha

Copy Editor: Safis Editing

Indexer: Subalakshmi Govindhan

Production Designer: Vijay Kamble

DevRel Marketing Coordinator: Nivedita Pandey and Anamika Singh

First published: August 2024

Production reference: 1020724

Published by Packt Publishing Ltd.
Grosvenor House
11 St Paul's Square
Birmingham
B3 1RB, UK

ISBN 978-1-83588-558-1

www.packtpub.com

Foreword

In the ever-evolving landscape of web development, mastering the latest technologies and integrating them seamlessly into full-stack projects is not just an advantage; it's a necessity. Having worked alongside *Mykyta Chernenko* in various capacities—from frontline coding at Quantum to collaborating on a startup idea—I have witnessed firsthand his deep technical expertise and innovative approach to problem-solving in the tech world.

Full-Stack Web Development with TypeScript 5 is Mykyta's latest endeavor to empower full-stack developers by sharing his wealth of knowledge. This book is a beacon for junior and mid-level software engineers who are eager to dive deep into the intricacies of modern web development using an array of cutting-edge tools including TypeScript, Bun, PostgreSQL, Svelte, and the pioneering OpenAI API.

What sets this book apart is not just its comprehensive coverage of both frontend and backend technologies but its practical, project-based approach. Through Mykyta's guidance, you will not only learn the theoretical aspects of each technology but also aspects of how to apply some of them in real-world scenarios, crafting robust and scalable web applications. The journey from understanding basic concepts to integrating AI features is laid out with clarity and precision, ensuring that every reader comes away with the ability to navigate and innovate within the tech industry.

As someone who has seen Mykyta's meticulous approach to both writing and coding, I am confident that this book will serve as a great resource for those looking to make advancements in their careers. Whether it's leveraging TypeScript's capabilities to enhance application reliability or integrating AI to create dynamic user experiences, readers will find *Full-Stack Web Development with TypeScript 5* an invaluable guide to becoming well-rounded, forward-thinking developers capable of tackling challenges for real business use cases.

Embark on this learning journey with Mykyta, and equip yourself with the knowledge to not just participate in the future of tech but to shape it.

Artem Korchunov

Senior Software Engineer, Pillsorted

About the reviewers

Kawtar Choubari is a software engineer specializing in JavaScript, TypeScript, React, and Next.js. She has been featured at Next.js Conf (2023) and React Conf (2024) with other community members and has spoken at major conferences, such as React Paris and Devoxx; she is passionate about sharing knowledge and making complex topics accessible.

Also, Kawtar mentors students, guiding their tech careers, creates tech videos part-time, and shares her conference insights in a vlog format on her YouTube channel (under her name). Dedicated to educating, she conducts workshops and actively engages on social media platforms, sharing her expertise and extending her impact with the tech community.

Kawtar's work is encapsulated by these three words: engineering, educating, entertaining.

Artem Korchunov is a product software engineer at Pillsorted, UK, where he blends a strong team-oriented approach with the ability to independently drive projects from inception to completion. Known for his project management and technical expertise in Terraform, AWS, .NET, and Node.js, Artem skillfully manages both the technical and business aspects of his work.

I am grateful to the team at Pillsorted for their support and the rich, collaborative experiences that drive mutual growth and innovation.

Joao Rodrigues is a senior typescript engineer at Acrontum Portugal.

Contributo

About the author

Mykyta Chernenko has over seven years of experience in technology, mainly in full-stack develo
with a focus on Python and TypeScript. He has also worked with Go, Kotlin, and Dart on va
projects. His technical contributions include working on the Azure integration for Nutanix's of
discovery project and a key engineering role at Factmata, overseeing engineering and infrastru
Currently, he holds a position as a senior consultant at KodeWorks. In addition to his technical
Mykyta has mentored over 20 professionals, sharing his knowledge and experience. He is al
author of *The Rational Software Engineer* and runs a blog with the same name on Hackernoon,
he writes about his insights in the field.

*A big thank you to my best friend, Artem, for working closely with me, offering valuable feedbac
always being there both as a friend and a technical reviewer. I also want to express my gratitude
everyone at Packt Publishing for their professionalism and for making the publishing process se
and positive. Your efforts have truly shaped this book.*

Table of Contents

Part 2: Backend Development with Bun and TypeScript

3

4

5

6

Advanced Backend Development – Security, Throttling, Caching, and Logging   87

Part 3: Integrating PostgreSQL for Data Management

7

PostgreSQL Basics, Storage, and Setup   105

8

Interacting with PostgreSQL Using Libraries   119

Part 5: Frontend Development with Svelte

12

13

14

15

Advanced Svelte Techniques   225

Preface

This book, *Full-Stack Web Development with TypeScript 5*, takes you on a journey into the robust and versatile world of TypeScript and will enable you to develop modern web applications from the ground up. With a focus on practical, real-world applications, this guide equips you with the necessary tools and techniques to master full-stack development using cutting-edge technologies such as Bun for the backend, Svelte for the frontend, PostgreSQL for database management, and the OpenAI API for AI integration. Whether you're looking to deepen your existing knowledge or venture into new aspects of web development, this book provides step-by-step instructions and a project-based learning approach that culminates in the creation of a full-featured chat application.

Who this book is for

This book is designed for junior to mid-level software engineers who have a basic understanding of JavaScript and web development principles. It is especially beneficial for those looking to enhance their skills in modern web development technologies and application design, focusing on both frontend and backend development with TypeScript.

What this book covers

Chapter 1, TypeScript Fundamentals, introduces TypeScript, explaining its evolution from JavaScript and highlighting its fundamental syntax. This chapter lays the foundations for understanding how TypeScript enhances web development capabilities.

Chapter 2, TypeScript Deep Dive – Typing, Generics, Classes, and Interfaces, explores advanced TypeScript features such as static typing, generics, and object-oriented programming concepts, which provide the tools we need to write more robust and maintainable code.

Chapter 3, Configuring a Backend Environment with Bun and Hono, guides you through setting up Bun as a modern JavaScript runtime optimized for TypeScript, focusing on configuring the environment to enhance backend development.

Chapter 4, Building Backend Infrastructure with Bun, Hono, and TypeScript, delves into creating a secure and efficient backend using Bun, Hono, and TypeScript, covering topics such as authentication systems, routing, and middleware integration.

Chapter 5, Improving Reliability – Testing and Validation, focuses on ensuring code quality and reliability in backend development, introducing techniques for data validation and testing strategies using Bun and TypeScript.

Chapter 6, Advanced Backend Development – Security, Throttling, Caching, and Logging, covers critical backend aspects such as security measures, caching strategies, and best practices for debugging and logging to build scalable applications.

Chapter 7, PostgreSQL Basics, Storage, and Setup, explores setting up a database using PostgreSQL within a Docker container to ensure reliable data persistence for web applications, detailing installation, schema construction, and CRUD operations.

Chapter 8, Interacting with PostgreSQL Using Libraries, advances our backend infrastructure by incorporating SQL interactions directly into our server code using the pg library. This chapter discusses efficient and secure data handling, connection strategies, and the importance of testing SQL integrations to bolster application reliability.

Chapter 9, Interacting with PostgreSQL Using Prisma ORM, transitions from direct SQL handling to using **Object-Relational Mapping** (**ORM**) for database interactions, focusing on using Prisma to streamline CRUD operations and manage schema migrations. This chapter covers how ORM integration can increase productivity, reduce errors, and make code clearer when working with PostgreSQL, as well as efficient migration management and testing ORM interactions.

Chapter 10, Basics of Integrating External APIs with TypeScript and Hono, concludes our backend setup by teaching how to integrate APIs using fetch for robust communications in TypeScript. This chapter focuses on implementing reliable API calls, including error handling, retries, and response validation, to ensure effective and resilient external service interactions.

Chapter 11, Setting Up and Configuring the OpenAI API for the Backend, details the process of integrating the OpenAI API into a TypeScript-based backend, focusing on configuration, security, and practical uses of AI models.

Chapter 12, Introduction to Svelte for Frontend Development, introduces Svelte, a modern framework for building reactive user interfaces, and highlights its key features and benefits over traditional frameworks.

Chapter 13, Setting up the Svelte Project, guides you through setting up a development environment for a Svelte-based application, exploring the configuration files and initial setup with tools such as Vite and SvelteKit.

Chapter 14, Svelte Chat Application Development, walks through the development of a chat application using Svelte, covering frontend aspects such as routing, state management, and UI design.

Chapter 15, Advanced Svelte Techniques, concludes the book with insights into maintaining and optimizing a Svelte and TypeScript application, covering testing, linting, accessibility, and internationalization strategies.

To get the most out of this book

Before diving into this book, it is recommended that you have a solid foundation in JavaScript and basic web development concepts. Familiarity with the essentials of both frontend and backend development will help you understand and implement the advanced techniques covered in this guide. This book assumes knowledge of web application structure and design patterns and aims to enhance your skills through practical application and real-world examples.

Software/hardware covered in the book	Operating system requirements
Bun 1.0	macOS or Linux
TypeScript 5.2	Windows, MacOS, or Linux
Hono 3.11	Windows, MacOS, or Linux
Svelte 4.2	Windows, MacOS, or Linux
Vite 5.0	Windows, MacOS, or Linux
Eslint 8.57	Windows, MacOS, or Linux
Svelte-routing 2.11	Windows, MacOS, or Linux
Prisma 5.7	Windows, MacOS, or Linux
Pg 8.11	Windows, MacOS, or Linux
Zod 3.22	Windows, MacOS, or Linux
Axios 1.6	Windows, MacOS, or Linux

If you are using the digital version of this book, we advise you to type the code yourself or access the code from the book's GitHub repository (a link is available in the next section). Doing so will help you avoid any potential errors related to the copying and pasting of code.

If you execute the code on Windows, be aware that Bun is in experimental mode and the instructions provided in the terminal must be used in a bash-like environment.

Download the example code files

You can download the example code files for this book from GitHub at `https://github.com/PacktPublishing/Full-Stack-Web-Development-with-TypeScript-5`. If there's an update to the code, it will be updated in the GitHub repository.

We also have other code bundles from our rich catalog of books and videos available at `https://github.com/PacktPublishing/`. Check them out!

Conventions used

There are a number of text conventions used throughout this book.

`Code in text`: Indicates code words in text, database table names, folder names, filenames, file extensions, pathnames, dummy URLs, user input, and Twitter handles. Here is an example: "For example, when `[]` is used in a numerical context, it is coerced to 0, and division of 4 to 0 results in `Infinity`. "

A block of code is set as follows:

```
const numbers: Array<number> = [1, 2]
const bigNumbers: number[] = [300, 400]
```

When we wish to draw your attention to a particular part of a code block, the relevant lines or items are set in bold:

```
export function createAuthApp(
  userResource: IDatabaseResource<DBUser, DBCreateUser>,
) {
  authApp.post(
    REGISTER_ROUTE,
    zValidator("json", registerSchema),
    async (c) => {
      const { email, password, name } = c.req.valid("json");
      ...
```

Any command-line input or output is written as follows:

```
$ npm install -g typescript
```

Bold: Indicates a new term, an important word, or words that you see onscreen. For instance, words in menus or dialog boxes appear in **bold**. Here is an example: "Turn on **Automatic ESlint config** if it was turned off."

> **Tips or important notes**
> Appear like this.

Get in touch

Feedback from our readers is always welcome.

General feedback: If you have questions about any aspect of this book, email us at `customercare@packtpub.com` and mention the book title in the subject of your message.

Errata: Although we have taken every care to ensure the accuracy of our content, mistakes do happen. If you have found a mistake in this book, we would be grateful if you would report this to us. Please visit `www.packtpub.com/support/errata` and fill in the form.

Piracy: If you come across any illegal copies of our works in any form on the internet, we would be grateful if you would provide us with the location address or website name. Please contact us at `copyright@packt.com` with a link to the material.

If you are interested in becoming an author: If there is a topic that you have expertise in and you are interested in either writing or contributing to a book, please visit `authors.packtpub.com`.

Share Your Thoughts

Once you've read *Full-Stack Web Development with TypeScript 5*, we'd love to hear your thoughts! Scan the QR code below to go straight to the Amazon review page for this book and share your feedback.

`https://packt.link/r/1835885594`

Your review is important to us and the tech community and will help us make sure we're delivering excellent quality content.

Download a free PDF copy of this book

Thanks for purchasing this book!

Do you like to read on the go but are unable to carry your print books everywhere?

Is your eBook purchase not compatible with the device of your choice?

Don't worry, now with every Packt book you get a DRM-free PDF version of that book at no cost.

Read anywhere, any place, on any device. Search, copy, and paste code from your favorite technical books directly into your application.

The perks don't stop there, you can get exclusive access to discounts, newsletters, and great free content in your inbox daily

Follow these simple steps to get the benefits:

1. Scan the QR code or visit the link below

https://packt.link/free-ebook/9781835885581

2. Submit your proof of purchase
3. That's it! We'll send your free PDF and other benefits to your email directly

Part 1:
Introduction to TypeScript

In this part, you will dive into the world of **TypeScript**, understanding its core principles and advanced features. This section sets a strong foundation in TypeScript, highlighting its evolution from JavaScript and demonstrating its power in enhancing code quality and maintainability for large-scale applications. You'll learn the basics as well as more complex concepts, such as generics and object-oriented programming, that are essential for robust web development.

This part includes the following chapters:

- *Chapter 1, TypeScript Fundamentals*
- *Chapter 2, TypeScript Deep Dive – Typing, Generics, Classes, and Interfaces*

1

TypeScript Fundamentals

Web development is constantly evolving, yet one constant remains: JavaScript as the primary language, a status it's likely to hold for the foreseeable future. Alongside JavaScript, an array of tools has surfaced, each contributing to the more efficient creation of full-stack applications. This book is dedicated to exploring these developments.

We'll focus on full-stack development, combining TypeScript with a variety of cutting-edge technologies needed to build complete applications through building a hands-on project. Our practical project involves constructing a chat application, similar in essence to the functionality of ChatGPT. It will involve the application of diverse frontend and backend technologies, along with database and API integrations.

This book is more than just a guide to coding an application from start to finish. It also aims to teach you about effective development patterns, which are versatile enough to be applied across various technologies of your choice. While it's assumed that you come with a basic understanding of JavaScript and fundamental web development concepts, every new topic and aspect introduced in the book will be thoroughly explained, ensuring a deep and comprehensive understanding of each subject.

In this book, we'll start by exploring TypeScript, now a key player in web development. Its main advantage is the use of types, which significantly improves the development process. We'll discuss TypeScript's history, compare it with JavaScript, and explore its strengths, particularly in enhancing code quality and reducing errors. We'll also delve into TypeScript's core concepts, such as syntax, types, interfaces, classes, and generics, providing a thorough understanding of how it works.

Then, we'll shift our focus to backend development, choosing Bun as our main technology. As of 2023, Bun is one of the most promising backend runtimes and is notable for its seamless integration with TypeScript. We'll cover the essentials of Bun, including setting up the environment, handling authentication, routing, middleware, data validation for requests, building REST APIs, using linters for code quality, debugging, logging, code structure, and effective testing strategies.

Next, we'll dive into using databases, specifically PostgreSQL, one of the most popular SQL-based databases with extensive features. We'll discuss data storage and how to optimize it. A key focus will be on using TypeScript as an interface for managing the data we store. We'll start with basic create, read, update, and delete operations, and then move on to using libraries and object-relational mappers to interact with PostgreSQL from TypeScript. This section will also cover integrating PostgreSQL into our backend infrastructure and optimizing its use.

Following that, we'll tackle API integration, which is crucial for our chat application. We'll be using the OpenAI GPT API for chat completions. While integrating and configuring this API for our backend, we'll cover broader topics, such as writing external service integrations effectively, ensuring the correctness of API responses, and incorporating API calls into our REST endpoints. Additionally, we'll discuss various scenarios where the OpenAI API and its models can be beneficial in web development.

The final section of this book focuses on frontend development using Svelte. Known for its simplicity, speed, and beginner-friendliness, Svelte is a standout choice for building web interfaces. We'll explore reactivity patterns common in single-page application frameworks and apply them to build our chat functionality using Svelte. This includes setting up the environment with TypeScript, understanding Svelte's core concepts and syntax, integrating the frontend with our REST API, and learning about components, routing, state management, and styling. Additionally, we'll delve into testing and debugging Svelte code and discuss best practices for maintaining and extending the code base.

By the end of this book, you'll gain more than just knowledge of the specific technologies covered. You'll develop a versatile mental framework for full-stack development. This framework will equip you with valuable concepts and practices that are applicable across any technologies and languages you will choose in web project development.

Let's kick off our journey with TypeScript. In this chapter, we'll dive into an introduction to TypeScript, covering its history, how it stands out from JavaScript, and its advantages. We'll also get acquainted with TypeScript's basic syntax, setting the stage for more advanced topics to follow. Here is the list of topics we are going to cover:

- Introduction to TypeScript and its evolution
- Key differences between TypeScript and JavaScript
- The advantages of using TypeScript in modern web development
- Basic syntax of TypeScript

Technical requirements

In this chapter, there's no need to install or run anything just yet. We're going to focus on the basics of TypeScript, so you can ease into it without any setup work.

All the code examples we discuss are available in this repository: `https://github.com/PacktPublishing/Full-Stack-Web-Development-with-TypeScript-5/tree/main/Chapter01`.

Introduction to TypeScript and its evolution

JavaScript's journey to market was remarkably quick. Brendan Eich created its first version in just 10 days, aiming to make web pages interactive through a straightforward scripting language. Over time, JavaScript gradually, then rapidly, became a dominant force in web development, while applications grew increasingly complex.

Many developers critique JavaScript's design as lacking elegance, often pointing out its seemingly inconsistent language decisions. It's true that JavaScript is notorious for its quirks. To illustrate this, let's look at two common examples:

- `4 / []`
- `0 == ""`

The first one evaluates to `Infinity`, while the second one becomes `True`. This happens because of the type coercion that happens when this code is executed. Type coercion is the process of automatically converting data types from one to another. For example, when `[]` is used in the numeric context, it is coerced to `0`, and division of `4` by `0` results in `Infinity`. In the second example, `""` is first converted to `0`, and then both sides of the equation become equal. Let's agree, this is pretty confusing.

Another big drawback people often point out is that JavaScript is weakly and dynamically typed. To fix this, Microsoft rolled out TypeScript in 2012. Since then, there's been a lot of changes in both JavaScript and TypeScript. For example, now in JavaScript, you can sidestep the issue in the first example by using the `===` comparison, which compares the values without type coercion. But the big win of TypeScript is still its static typing, which would not allow such comparison in the first place as a `number` cannot be equal to a `string` by a type definition, so the comparison doesn't make sense in type logic. This is far from all of the benefits strict typing provides to us. We will explore more of its benefits further in this chapter and in the next one.

Let's take a closer look at what sets TypeScript apart from JavaScript.

Key differences between TypeScript and JavaScript

TypeScript is a superset of JavaScript, meaning everything you can do in JavaScript, you can also do in TypeScript. It works in all the same places as JavaScript – browsers, Node.js, Bun, and so on. Before the code execution, TypeScript is first transpiled into JavaScript. So, what actually runs is plain JavaScript. TypeScript exists only during development.

This means all your JavaScript knowledge is still valuable in TypeScript. But TypeScript adds a bunch of cool stuff:

- **Type annotations**: TypeScript allows type annotations, where you can explicitly declare variable types. JavaScript does not support this natively.

- **Interfaces**: TypeScript introduces interfaces, a way to define custom types and ensure objects conform to specific structures. JavaScript lacks this feature.

- **Access Modifiers**: TypeScript supports access modifiers such as `private`, `protected`, and `public`, for controlling access to class members. JavaScript does not have this feature built-in.

- **Enums**: TypeScript provides enums, a feature for defining a set of named constants. This is not available in standard JavaScript.

- **Namespaces and modules**: TypeScript offers namespaces for grouping code and avoiding global scope pollution, and it has robust support for ES6 modules. JavaScript primarily relies on ES6 modules.

- **Advanced types**: TypeScript has advanced type features, such as generics, union types, and tuple types, allowing more precise type definitions and manipulation. JavaScript does not have these advanced type features.

- **Tooling and compilation**: TypeScript requires a compilation step to transpile TypeScript code to JavaScript, which can be integrated into build processes. JavaScript can be run directly in browsers and Node.js without this compilation step.

As we develop our chat application, we'll dive deeper into these TypeScript features and put them to practical use. But before that, let's discuss the advantages and disadvantages of using TypeScript.

The advantages of using TypeScript in modern web development

Using TypeScript comes with a few advantages, mostly of which come directly from using types:

- **Type safety**: TypeScript's static typing helps catch errors at compile time, long before the code is executed. This leads to fewer runtime errors and more robust, reliable code.

- **Improved tooling**: The static type system allows for better tooling support like auto-completion, navigation, and refactoring tools, making the development process more efficient.

- **Easier maintenance**: For large code bases, TypeScript's type system makes the code easier to understand and maintain. Changes can be made with greater confidence, reducing the likelihood of introducing bugs.

- **Better documentation**: The type annotations serve as a form of documentation, making it easier for new developers to understand what the code is doing.

At the same time, as with every technology, TypeScript is not all that shiny; there are disadvantages as well:

- **Learning curve**: For developers familiar with JavaScript, learning TypeScript introduces an additional layer of complexity due to its static typing and other advanced features.

- **Compilation step**: TypeScript code must be compiled to JavaScript before it can be executed. This adds an extra step to the development process and can complicate build and deployment pipelines. The configuration can become even more complex than plain JavaScript, and with any additional step in the configuration, it is more likely to break.

- **Potentially verbose**: Type annotations and interfaces can make TypeScript code more verbose than JavaScript. This can lead to longer, more complex code, which might be seen as a downside for simple projects.

- **Community and ecosystem adjustments**: While TypeScript is widely adopted, some libraries and third-party tools may still have better support for JavaScript. This means developers might need to invest additional effort to find or create TypeScript type definitions for existing JavaScript libraries.

- **Integration with JavaScript infrastructure**: Since TypeScript is converted to JavaScript for runtime, the type information is lost during execution. This can make using third-party TypeScript libraries a bit inconvenient. When exploring a function from a library, you often end up in the `*.d.ts` files, which define only the type structure of the functions and not the definitions.

Overall, TypeScript isn't flawless, but the type safety it offers leads to more reliable and error-free code. This benefit is often more crucial for larger projects than the drawbacks TypeScript might present. With the understanding you have now, you're well prepared to move on to something more practical – the basic syntax of TypeScript.

Basic syntax of TypeScript

Let's start with two main aspects of TypeScript – simple types and interfaces.

Simple types

JavaScript has types, but they are implicit and not strictly enforced. In TypeScript, you'll find most types are explicit, and you're required to declare them. Let's declare one:

```
let messageText: string = "my first chat message"
```

The difference you can see with the ordinary JavaScript is `: string`. This part defines the time of the variable that we are going to use. Here, we've clearly stated that `messageText` is a string. If we try to assign a value of a different type, like a number, we'll get an error:

```
messageText = 5
```

This line will trigger an error since 5 is not a string.

Another benefit is when the compiler knows the variable is a string, your IDE will suggest actual functions that exist on the string type, which is quite useful. Similarly, we can use other basic types to annotate variables.

There are a few more basic types to know:

- Number and `boolean`: The first two, `number` and `boolean`, are also primitives, like the string type.

- `Array<T>`: `Array<T>` is a generic type, a concept we'll explore in *Chapter 2*, but it's essentially for declaring arrays with a specific type of elements. This can also be written using `[]`.

  Let's declare two arrays of numbers to demonstrate:

  ```
  const numbers: Array<number> = [1, 2]
  const bigNumbers: number[] = [300, 400]
  ```

  In this code block, we define two variables – `numbers` and `bigNumbers`. They define their types differently, but essentially `Array<number>` and `number[]` mean the same – an array of numbers.

- any: The last type, any, represents any type. Using any tells TypeScript to stop checking the type. It's useful for converting JavaScript to TypeScript, but it basically negates all the benefits TypeScript offers.

I encourage you to experiment with these basic types in your IDE for a better grasp. We'll also be using them extensively as we build the app, so don't worry if it doesn't all click right away. Now, let's briefly touch on interfaces.

Interfaces

An **interface** in TypeScript defines the structure of an object, specifying what fields it should have and the types of those fields. Let's create a simple interface for a chat object and then define a few instances of the Chat type. We will complete the code piece with a function called displayChat, which accepts a parameter of type Chat and logs its details:

```
interface Chat {
    name: string;
    model: string;
}

const foodChat: Chat = { name: 'food recipes exploration', model:
'gpt-4' };
const typescriptChat: Chat = {name: 'typescript teacher', model: 'gpt-
```

```
3.5-turbo'};

function displayChat(chat: Chat) {
    console.log(`Chat: ${chat.name}, Model: ${chat.model}`);
}
```

First, we define an interface called `Chat` that contains two fields `name` and `model` of type `string`. By doing so, we create our own type that we can use in variable type annotations as we did with basic types in the beginning of the *Simple types* section. When we use it on a variable, it essentially means that the object that we assign to the variable has to follow the structure of the interface.

Then, we define the two variables, `foodChat` and `typescriptChat`, both of which satisfy the `Chat` interface. They hold different data, but both are of type `Chat`. The `displayChat` function accepts any parameter that satisfies the `Chat` interface, meaning that it is an object with `name` and `model` fields of the `string` type.

Interfaces also give an error if you try to access a property that doesn't exist on the object. They can be extended, have optional properties, and include function definitions. We'll explore all these features in the upcoming chapters.

Summary

In this chapter, we've covered a brief introduction to TypeScript – how it relates to JavaScript, its advantages and disadvantages, its history, and some basic functionality. With this information in hand, we're ready to dive deeper into TypeScript. In the next chapter, we are going to cover more advanced aspects of TypeScript: **generics**, **unions**, **classes**, and other useful functionality of its type system.

2

TypeScript Deep Dive – Typing, Generics, Classes, and Interfaces

We covered the very basics of TypeScript in *Chapter 1*. Now, let's go further, and learn more advanced features that are going to be essential to develop any real-world application using TypeScript. In this chapter, we are going to build some foundational types that we are going to use in our backend and frontend applications, and along the way, we will learn about the following concepts:

- Advanced typing techniques
- Creating types from other types
- Interface and **object-oriented programming (OOP)** features
- Generics
- Promises

We will start with advanced typing techniques, which will include narrowing, `null` types, and function types.

Technical requirements

In this chapter, we will need to install TypeScript. I encourage you to experiment with the code we present here and play around with your own examples to understand the mechanics of the techniques I'm going to present, better.

You can install TypeScript globally by using this command in your terminal of choice:

```
$ npm install -g typescript
```

Now, you have the `tsc` command-line tool installed system-wide. Another thing that we need to do when we create a project is to provide TypeScript configuration for the project. Let's create a `Chapter02` folder and create a `tsconfig.json` file inside of it. Put the following configuration into it:

tsconfig.json

```
{
  "compilerOptions": {
    "module": "es2022",
    "target": "es2017",
    "strictNullChecks": true
  },
  "includes": [
    "main.ts"
  ]
}
```

This configuration is a minimalistic configuration required to transpile our TypeScript to JavaScript and it also requires strict checks for `null` values, which we mentioned in the previous chapter.

To transpile the code, you can write this command in the terminal inside the `Chapter02` folder:

```
$ tsc -p tsconfig.json
```

We are going to put all the code mentioned in this chapter in the `main.ts` file.

The project code we discuss here is also available in this repository:

`https://github.com/PacktPublishing/Full-Stack-Web-Development-with-TypeScript-5/tree/main/Chapter02`

Advanced typing techniques

We will start our chapter with more advanced typing techniques. We will explore key concepts, including **narrowing**, **null types**, **function types**, and a suite of **utility types** such as `Partial`, `Readonly`, `Required`, `Pick`, `Record`, and `Omit`. The first technique we are going to look at will be narrowing, which is going to help limit what a type can be.

Narrowing

When we talk about narrowing, we refer to the process of moving from a less precise type to a more precise type. For instance, a variable that initially has a type of `any` or `unknown` can be narrowed down to more specific types such as string, number, or custom types.

Let's delve into a practical example to illustrate how narrowing works in a real-world scenario. In the following code, we'll discuss a `narrowToNumber` function and a `getChatMessagesWithNarrowing` function, both of which utilize narrowing:

```
function narrowToNumber(value: unknown): number {
    if (typeof value !== 'number') {
        throw new Error('Value is not a number');
    }
    return value;
}
```

This function takes a parameter of type `unknown` and aims to ensure that this parameter is indeed a number. The usage of `typeof` in the `if` statement is a classic example of type guarding, a form of narrowing. If the value is not of type `number`, an error is thrown; otherwise, the function safely returns the value, now assured to be a number. This is a classic example of runtime type checking in TypeScript.

Let's move to an example of how we can get messages on our backend. You don't need to know a lot of the specifics of the backend-related code happening in this example, but I will explain the gist and the parts that are relevant:

```
async function getChatMessagesWithNarrowing(chatId: unknown, req: {
authorization: string }) {
    const authToken = req.authorization;
    const numberChatId = narrowToNumber(chatId);
    const messages = await chatService.getChatMessages(numberChatId,
authToken);

    if (messages !== null) {
        messages.map((message) => {
            console.log(`Message ID: ${message.id}, Feedback:
${message.feedback?.trim() ?? "no feedback"}`);
        });
        return {success: true, messages}
    } else {
        return {success: false, message: 'Chat not found or access
denied'}
    }
}
```

Here, `chatId` is passed to the function with an unknown type. However, in our scenario, `chatId` is expected to be a number. This is where the `narrowToNumber` function becomes useful. By applying it to `chatId`, we are narrowing the type from a broad unknown type to a more specific type of number type. Further down in the code, we encounter the following instance of narrowing:

```
if (messages !== null) {
    ....
}
```

This line is a subtle but vital example of narrowing. Here, we check if `messages` is not `null` before proceeding. This act effectively narrows the type of `messages` from possibly `null` or an array of `messages` to definitively an array of `messages`. This ensures that the operations inside the `if` block are safe and won't result in a `TypeError` error due to attempting to access properties on `null`.

Now, let's look into another type feature – `null` types – in the next section.

null types

In JavaScript and TypeScript, `null` is a primitive value that represents the intentional absence of any object value. When TypeScript's strict `null` checking is enabled (which is highly recommended), variables must be explicitly typed to include `null` or `undefined` if they are intended to hold an empty value. This contrasts with JavaScript, where variables can implicitly be `null` or `undefined` without such strict type distinctions, often leading to unintended errors if not carefully managed. This forces developers to consciously handle `null` cases, leading to more robust and error-resistant code. We've seen it in action already in the previous section, but let's look at it closer:

```
if (messages !== null) {
    // ...
}
```

Here, the `messages !== null` `null` check is a direct application of handling `null` types. In this context, `messages` is expected to be an array or `null`. The check ensures that the following code only runs if `messages` is indeed an array and not `null`. This is a simple yet effective way to guard against `null`-related errors.

Next, let's examine another snippet here:

```
console.log(`Message ID: ${message.id}, Feedback: ${message.feedback?.
trim() ?? "no feedback"}`);
```

In this line, the use of the `?` optional chaining operator and the `??` nullish coalescing operator provides a powerful combination for dealing with `null` values, which helps to prevent runtime errors where we would expect a value but got `null`. The expression `message.feedback?.trim()` expression will only attempt to call `trim()` if feedback is not `null` or `undefined`, thus avoiding a potential runtime error. If feedback is nullish (that is, `null` or `undefined`), the nullish coalescing operator takes over, providing a `"no feedback"` fallback value.

`null` types in TypeScript are not just a feature of the language; they represent a mindset shift toward safer, more predictable coding practices, as `Uncaught TypeError: Cannot read properties of null` is an error that is very commonly met in JavaScript but is completely gone from TypeScript.

Now, in the next section, let's look at function types.

Function types

A function type allows you to specify the exact form a function should take: the types of its input arguments and its return type. This feature is helpful when you pass functions around the code as arguments or if you create a constant of a function type.

Let's look at an example of how we can use a function type when we log the details of every message in our code:

```
type MapCallback = (message: IMessage) => void;
const logMessage: MapCallback = (message) => {
    console.log(`Message ID: ${message.id}`);
};
messages.map((message: IMessage) => {
    logMessage(message);
});
```

Here, `MapCallback` is a function type definition. It tells TypeScript that any function with this type should take one argument, `message`, which is of type `IMessage`, and it should return nothing (`void`). This function type becomes a template for creating functions with this specific structure.

`logMessage` is a function that's explicitly declared to be of type `MapCallback`. This means `logMessage` must match the structure defined by `MapCallback` – it takes an `IMessage` object as an argument and does not return anything. TypeScript will enforce this structure, ensuring that `logMessage` is used correctly according to the defined function type.

Finally, `logMessage` is used within the map function. Each item in messages (assumed to be an array of `IMessage` objects) is passed to `logMessage`, which adheres to the structure defined by `MapCallback`. This ensures that the function is applied correctly to each `message` type and that we can only pass an argument of type `IMessage` to the `logMessage` function.

With this, we are ready to move to the next topic, creating types from other types, which will help us change existing types to get something new.

Creating types from other types

In this section, we are going to cover a few techniques to adapt one type to another, such as **utility types**, **union types**, and **intersection types**. We will start by showing some useful utility functions for this job in TypeScript.

Utility types

Utility types are a set of types provided by the language to transform existing types into new, modified versions. They offer a convenient way to alter properties of a type, making them optional, read-only, or excluding them, among other transformations. Let's cover a few of them here:

- `Pick<Type, Keys>`: Pick constructs a new type by selecting a subset of properties from an existing type. `Pick` is used when you need a type with only specific properties from a parent type. This is useful for creating more focused and less bulky types. The following is an example of using `Pick`:

```
interface IUser {
  id: number;
  name: string;
  email: string;
}
type UserPreview = Pick<IUser, 'id' | 'name'>;
const userPreview: UserPreview = {id: '1', name: 'John'};
```

  Here, `UserPreview` contains only `id` and `name` from `IUser`.

- `Record<Keys, Type>`: Record generates a type with a set of keys and assigns a uniform type to these keys' values. It is ideal for creating objects where keys have a common value type, often used for mapping or lookup purposes. Here is an example of using `Record`:

```
type UserNamesById = Record<UserId, string>;
const userNamesById: UserNamesById = {'1': 'John', '2':
'Alice'};
```

  `UserNamesById` is a dictionary object mapping `UserId` keys to string names.

- `Partial<Type>`: Partial turns all properties of a given type into optional properties. `Partial` is useful in situations such as updating parts of an object, where you don't need to provide all properties. It provides flexibility in object creation. Here is how you would use `Partial` in code:

```
type PartialIUser = Partial<IUser>;
const partialUser: PartialIUser = {id: '1'};
```

  `PartialIUser` allows any combination of `IUser` properties, including incomplete objects, because all the properties are optional.

- `Required<Type>`: `Required` transforms all optional properties of a type into required ones. `Required` is the opposite of `Partial`. It enforces that all properties of the type must be provided, ensuring complete object definitions. The following is an example of using `Required`:

```
type RequiredIUser = Required<PartialIUser>;
const requiredUser: RequiredIUser = {id: '1', name: 'John',
email: 'john@example.com'};
```

  `RequiredIUser` mandates that all properties, even those optional in `PartialIUser`, must be present.

- `Omit<Type, Keys>`: `Omit` creates a new type by omitting specified properties from an existing type. `Omit` is useful for creating a type that excludes certain properties from another type, which is particularly helpful in excluding sensitive or unnecessary properties. An example of `Omit` is as follows:

```
type UserWithoutEmail = Omit<IUser, 'email'>;
const userWithoutEmail: UserWithoutEmail = {id: '2', name:
'Alice'};
```

  `UserWithoutEmail` is a type similar to `IUser` but without the email property.

- `Readonly<Type>`: `Readonly` makes all properties of a type immutable post-creation. `Readonly` is used to create types where object properties shouldn't be changed after the object is created, which is important for maintaining integrity in certain objects:

```
type ReadonlyIUser = Readonly<IUser>;
const user: ReadonlyIUser = {id: '1', name: 'John', email:
'john@example.com'};
```

  `ReadonlyIUser` guarantees that once an `IUser` object is created, its properties cannot be altered, as we cannot assign a new field to it afterward.

With this covered, let's see how we can combine types together with union types, in the next section.

Union types

Union types, represented by a pipe sign (|), allow a variable to hold values that are a combination of two or more types, offering flexibility in defining types that can accept multiple, specific types of values. They're essential in scenarios where a variable or function return type isn't confined to a single type.

Let's apply this concept to the provided examples and create a type for our `IMessage` interface:

```
type MessageType = "user" | "system";
```

Here, `MessageType` is a union type, meaning it can hold either a `"user"` or `"system"` value. Now, let's add it to our interface, as shown here:

```
interface IMessage {
    type: MessageType;
    // other properties
}
```

In the `IMessage` interface, the type property must be either `"user"` or `"system"`, adhering to the `MessageType` union. However, union types don't have to be primitive values. Let's now look at how we can handle returning a union type from a function. The `getChatFromDb` function here illustrates a common use case for union types in function return values:

```
type DbChatSuccessResponse = {
    success: true;
    data: IChat;
};

type DbChatErrorResponse = {
    success: false;
    error: string;
};

function getChatFromDb(chatId: string): DbChatSuccessResponse |
DbChatErrorResponse {
    const findChatById = (_: string) => ({} as IChat)
    const chat = findChatById(chatId);
    if (chat) {
        return {
            success: true,
            data: chat,
        };
    } else {
        return {
            success: false,
            error: "Chat not found in the database",
        };
    }
}
```

getChatFromDb can return either DbChatSuccessResponse, which represents a successful operation, or DbChatErrorResponse, which has properties for a failed response. This approach is useful for error handling and data fetching, where the outcome might differ significantly. Now, let's handle the union type result of the function with narrowing in the following code:

```
const dbResponse = getChatFromDb("chat123");
if (dbResponse.success === true) {
    console.log("Chat data:", dbResponse.data);
} else {
    console.error("Error:", dbResponse.error);
}
```

In this snippet, the response from getChatFromDb is either a success or an error object. The if dbResponse.success === true check effectively distinguishes between these two possible return types. If success is true, TypeScript understands that dbResponse conforms to DbChatSuccessResponse and allows access to dbResponse.data. Otherwise, it treats dbResponse as DbChatErrorResponse, exposing the error property.

Let's next look at a resembling technique called type intersections.

Type intersections

Type intersection is a feature that creates a new type that includes all properties of combined types. It's symbolized by an ampersand (&) and is particularly useful for composing complex types from simpler ones. In the following code, we are going to create a type for a database chat entity that must also have an id value:

```
type IDBEntityWithId = {
    id: number;
};

type IChatEntity = {
    name: string;
};

type IChatEntityWithId = IDBEntityWithId & IChatEntity;

const chatEntity: IChatEntityWithId = {
    id: 1,
    name: "Typescript tuitor",
};
```

Here, `chatEntity` is declared with the `IChatEntityWithId` type, so it must include both `id` from `IDBEntityWithId` and `name` from `IChatEntity`. This illustrates how intersection types enforce the presence of all properties from combined types. We can also achieve similar functionality by extending interfaces, which we are going to show in the following topic.

With this, we've covered the essential part of advanced typing techniques, and we are ready to talk more about how to reuse and write extensible code with **OOP** and **interfaces**.

Interface and OOP features

Both OOP and interfaces serve a few very important purposes – they help to clearly define the structure of objects passed around, reuse code, and write extensible functionality. We've looked at interfaces before, but now, let's talk about how we can extend their definitions.

Interfaces

Extending interfaces allows for the creation of new interfaces that inherit properties from existing ones, thereby enhancing reusability and organization. Let's draw on an example we used before, but now using extending interfaces there:

```
interface IMessageWithType extends IMessage {
    type: MessageType;
}

const userMessage: IMessageWithType = {
    id: 10,
    chatId: 2,
    userId: 1,
    content: "Hello, world!",
    createdAt: new Date(),
    type: "user",
};
```

The `IMessageWithType` interface extends the `IMessage` interface, which means it includes all properties from `IMessage` plus any additional properties defined in `IMessageWithType`. Here, `IMessageWithType` adds the `type` property, of type `MessageType`, to the existing structure. When creating a `userMessage` object of type `IMessageWithType`, it's required to include all properties from both `IMessage` and `IMessageWithType`.

Now, let's look at OOP functionalities that exist in TypeScript.

OOP functionalities

OOP is a programming paradigm that uses objects and classes to create models based on the real world. TypeScript embraces core OOP principles, allowing developers to use polymorphism, abstraction, inheritance, and encapsulation using native syntax. Let's look at these functionalities in detail here:

- **Polymorphism:** This allows objects of subclasses to be treated as objects of a common superclass. It's about creating a structure where a function can utilize subclasses of the superclass interchangeably. Polymorphism is primarily achieved through interfaces and abstract classes. By defining a common interface or an abstract class, TypeScript allows different classes to implement the same structure or methods, thus enabling functions to work with objects of these different classes as if they were working with the base class.

- **Abstraction:** TypeScript uses abstract classes and interfaces to implement abstraction. These constructs allow you to define a standard template or contract that other classes can implement, encapsulating complex logic and exposing only the necessary parts.

- **Inheritance:** This is a mechanism where a new class extends (inherits from) an existing class, allowing for the reuse of code and creating a hierarchical relationship between classes. In TypeScript, inheritance is implemented using the `extends` keyword. A class can extend another class, inheriting its properties and methods.

- **Encapsulation:** This involves bundling data and methods that operate on the data within one unit, often a class, and restricting access to some of the object's components, which ensures data integrity. Encapsulation in TypeScript is achieved through access modifiers such as `public`, `private`, and `protected`. These modifiers control the visibility and accessibility of class members, ensuring that internal details of a class are hidden and only exposed through a defined interface. Here is a description of access modifiers:

  - `public`: This is the default access level for class members. Members declared as `public` can be accessed from anywhere, meaning there's no restriction on access. This includes access from within the class itself, from instances of the class, and from subclasses.

  - `private`: Members declared as `private` can only be accessed from within the class itself. They are not accessible from instances of the class or from subclasses. This access level is used to hide the internal state and functionality of the class from the outside, enforcing encapsulation.

  - `protected`: Members declared as `protected` can be accessed from within the class and also from subclasses. However, they are not accessible from instances of the class (unless through methods defined within the class or subclass). This allows for a more controlled form of accessibility, useful for cases where the subclass needs more intimate knowledge of the superclass without exposing members to the general public.

Let's look at the following example that combines these techniques. We will define an `AbstractDatabaseResource` abstract class with common methods and an `abstract` method. An `InMemoryChatResource` concrete class will extend this abstract class and provide specific implementations for storing chat in memory:

```typescript
abstract class AbstractDatabaseResource {

    constructor(protected resourceName: string) {
    }

    protected logResource(resource: { id: number }): void {
        console.log(`[${this.resourceName}] Resource logged:`,
resource);
    }

    abstract get(id: number): { id: number } | null

    abstract getAll(): { id: number }[]

    abstract addResource(resource: { id: number }): void;
}

const inMemoryChatResource = new InMemoryChatResource();

const chat1: IChat = {
    id: 1,
    ownerId: 2,
    messages: []
};

inMemoryChatResource.addResource(chat1);
const retrievedChat1 = inMemoryChatResource.get(1);
```

Let's discuss various techniques we have used here:

- **Abstraction**: The `AbstractDatabaseResource` class provides external methods that are used to manage database elements (methods such as `get`, `getAll`, and `addResource`), while it hides the complexity of the specific implementation. Users of the `AbstractDatabaseResource` class need only be concerned with the interface – what methods are available and what parameters they accept – not how these methods are implemented. This **separation of concerns** (**SoC**) makes the system easier to understand and use.

- **Inheritance**: `InMemoryChatResource` is a concrete class that extends `AbstractDatabaseResource`. This means it inherits its properties and methods, but also provides specific implementations for the abstract methods defined in the base class.

- **Encapsulation**: In `InMemoryChatResource`, the `resources` array is marked as `private`, meaning it can't be accessed directly from outside the class. This encapsulation ensures that the internal representation of chat resources is hidden from external use. The `logResource` method in the abstract class is marked as `protected`, allowing it to be accessed within the class and its subclasses, but not outside.

- **Polymorphism**: While `InMemoryChatResource` has different implementations of the methods (such as `addResource` and `get`), they can be used interchangeably in contexts where an `AbstractDatabaseResource` class is expected. This is polymorphism, where objects of different classes can be treated as objects of a common superclass.

- **Instantiation and use**: We create an instance of `InMemoryChatResource` and use it to add and retrieve chat data. Despite the specific underlying implementation (in-memory array), the code only relies on the use of shared abstract class structure definition to know the methods and properties it can retrieve.

It is important to use interfaces and classes appropriately. As a rule of thumb, leverage interfaces for defining contracts and shapes of data, ensuring consistency across implementations and facilitating easy refactoring. Use classes to encapsulate data and behavior, taking advantage of inheritance and polymorphism to promote code reuse and maintainability, while keeping class definitions focused and avoiding overly complex inheritance hierarchies. Interfaces are good for defining the abstract structure that you are going to operate on in a function parameter, and classes are great for encapsulating complexity and provide only a simple and straightforward way to interact with the classes' logic.

We can now work on the class code we introduced to cover two additional concepts: **generics** and **promises**.

Generics

Generics in TypeScript are a tool for creating reusable components that can work with multiple types rather than a single one. This allows for better component abstraction while maintaining type safety as well. It also helps to write more flexible code that can adapt to different types. Let's start with a basic example given here:

```
function printValue<T>(value: T): void {
    console.log(value);
}

printValue<number>(123);
printValue<string>("Hello");
```

\<T\> is a generic type parameter that allows this function to accept any type of value. When you call `printValue<number>(123)` and `printValue<string>("Hello")`, TypeScript treats T as `number` and `string`, respectively. This demonstrates how generics provide flexibility without losing the benefits of type checking.

We also can improve `InMemoryChatResource`, which we defined before. It works fine now, but what if we also need an in-memory implementation of `InMemoryUserResource` for `IUser` in addition to `IChat`? They basically have the same functionality; the only difference between them is the type of resource they handle. Generics can be helpful here. To illustrate that, let's define a generic `GenericsInMemoryResource` class and create `chat` and `user` instances of it here:

```
class GenericsInMemoryResource<T extends { id: number }> extends
AbstractDatabaseResource {
    private resources: T[] = [];

    constructor(resourceName: string) {
        super(resourceName);
    }

    get(id: number): T | null {
        const resource = this.resources.find((item) => item.id ===
id);
        return resource ? {...resource} : null;
    }

    getAll(): T[] {
        return [...this.resources];
    }

    addResource(resource: T): void {
        this.resources.push(resource);
        this.logResource(resource);
    }
}

const userInMemoryResource = new
GenericsInMemoryResource<IUser>('user')
const chatInMemoryResource = new
GenericsInMemoryResource<IChat>('chat')

userInMemoryResource.addResource({id: 1, name: 'Admin', email: 'admin@
admin.com'});
chatInMemoryResource.addResource({id: 10, ownerId:
userInMemoryResource.get(1)!.id, messages: []});
```

The `<T extends { id: number }>` syntax means T can be any type, but it must have an `id` property of type number. This ensures type safety while allowing for different resource types. Unlike `InMemoryChatResource`, which was limited to handling `IChat` objects, `GenericsInMemoryResource` can handle any type that meets the constraint. This eliminates the need for separate classes for each resource type, such as chats or users.

Instances such as `userInMemoryResource` and `chatInMemoryResource` demonstrate this using `GenericsInMemoryResource` with `IUser` and `IChat` types. Methods such as `addResource` and `get` work with the generic type T, making them adaptable to the specific type of resource being used.

The original `InMemoryChatResource` class is limited to handling chat data, while `GenericsInMemoryResource` is more flexible and scalable, capable of handling various types of resources. This leads to cleaner, more maintainable code, as you don't need to create a new class for each resource type.

Generics significantly reduce code duplication by enabling a single class to manage multiple data types. This approach simplifies maintenance and enhances the adaptability of our code base.

Now, let's briefly talk about how to handle promises with types.

Promises

In TypeScript, when you declare a promise, you can use generics to indicate the type of data the promise will eventually return. This type indication ensures that the resolved value is consistent with expectations and allows TypeScript to provide relevant type checking and autocompletion. Let's declare a `fetchData` function here that will imitate a network request with `setTimeout` and provide the type for its return value using the `Promise` type:

```
function fetchData(): Promise<string> {
    return new Promise((resolve) => {
        setTimeout(() => resolve("Data Fetched"), 1000);
    });
}
```

`fetchData` is a function returning a `Promise<string>` instance. This means the promise, when resolved, will return a string. Inside the promise, we simulate fetching data with `setTimeout` and resolve it with a `string` value, adhering to the declared return type `string`. This explicit typing of the promise's resolved value ensures that any consumer of `fetchData` can expect a string.

With this, we have covered most aspects of TypeScript that we will need to build our chat application.

Summary

In this chapter, we've delved into advanced TypeScript features, gaining a deeper understanding of concepts such as narrowing, `null` types, function types, and utility types, which are fundamental for developing robust web applications. This knowledge equips us with the tools to write safer, more predictable, and efficient code, reducing common JavaScript pitfalls. Moving forward, the next chapter will mark the beginning of our hands-on journey, where we'll start building our application by configuring our backend runtime environment with **Bun**, laying the groundwork for our full stack TypeScript project.

Part 2:
Backend Development with Bun and TypeScript

This part focuses on backend development using **Bun**, a modern JavaScript runtime, and TypeScript. You will learn how to set up the Bun environment and build a secure, scalable, and robust backend infrastructure. This part covers everything from configuring your development environment to implementing complex backend logic, including authentication, routing, and middleware, as well as best practices for testing and validation.

This part includes the following chapters:

- *Chapter 3, Configuring a Backend Environment with Bun and Hono*
- *Chapter 4, Building Backend Infrastructure with Bun, Hono, and TypeScript*
- *Chapter 5, Improving Reliability – Testing and Validation*
- *Chapter 6, Advanced Backend Development – Security, Throttling, Caching, and Logging*

3

Configuring a Backend Environment with Bun and Hono

In this chapter, we will start with understanding what the **Bun JavaScript runtime** is and why it's becoming a popular JavaScript runtime among developers. We'll then introduce a library for developing backends called Hono, which provides a simple yet powerful way to build web applications. You'll set up your project, including installing Bun and Hono, learn about some useful code structures, and configure essential tools such as middleware, environment files, Prettier, and ESLint.

By the end of this chapter, you'll have a strong grasp of Bun's environment and will be able to set up your own backend projects. All of this is going to be essential for you to be able to develop functional and complete backend applications and build the foundation for our chat app. You will also learn how to improve the readability and configurability of our code and make it easier to develop.

In this chapter, we're going to cover the following main topics:

- Introducing Bun
- Introducing Hono
- Setting up your project
- Adding linting and formatting
- Adding middleware
- Handling environment variables
- Discussing the project's structure

Technical requirements

First, we need to install Bun.

For Linux and MacOS, use the following command:

```
$ curl -fsSL https://bun.sh/install | bash
```

Bun has limited experimental support for Windows, so it's recommended to use **Windows Subsystem for Linux (WSL)**, but you can still install it using the following command:

```
$ powershell -c "irm bun.sh/install.ps1|iex"
```

If you get a `command not found` error, you need to add the `C:/Users/currentUser/.bun` folder and `C:/Users/currentUser/.bun/bin` to the Windows PATH system variable.

This is enough to get started, and we will install more libraries as we go. All the code examples we discuss are available in the GitHub repository:

```
https://github.com/PacktPublishing/Full-Stack-Web-Development-with-
TypeScript-5/tree/main/Chapter03
```

Introducing Bun

Bun is a JavaScript runtime and is an alternative to Node.js and Deno. It's rapidly gaining traction due to its exceptional performance and ease of use. It has native support for TypeScript and a lot of other benefits, including the following:

- **High-performance runtime**: Bun is built on the JavaScriptCore engine used in Safari, which is known for its speed. This and other optimizations make Bun an incredibly fast runtime, offering a performance boost, sometimes handling 5 or 10 times as much load as Node.js can.

- **Built-in bundling and transpiling**: Unlike traditional runtimes that require external tools for these tasks, Bun comes with built-in capabilities for bundling and transpiling. So, you only need Bun to interpret your code and transpile your code, so we don't need configuring tools such as webpack, esbuild, or rollup.

- **TypeScript and JSX support**: Bun provides first-class support for TypeScript and JSX out of the box.

- **Efficient package management**: Bun's package manager is designed to be faster and more efficient than traditional package managers such as npm or yarn, and is from 2 to 10 times quicker in different scenarios.

- **Node.js compatible**: Bun is designed to be a drop-in replacement for Node.js. Bun natively implements most of the essential Node.js and Web APIs. So, you can use packages developed for Node.js in Bun and even replace Node.js with Bun in your existing project. It doesn't work in 100% of cases, but the compatibility is fairly high, and most projects will run as they are without needing to change much.

- **Useful built-in libraries and tools**: Bun provides some important tooling that is needed to write most real projects out of the box, such as test runners, environment variable handlers, and password generators. This is going to save us quite some time because, otherwise, we would need to introduce external libraries for these functions anyway.

If you also want to see more of the comparison between Node.js, Deno, and Bun, you can visit `https://bun.sh/` to see the charts that compare them.

Bun will be quite helpful in developing our backend chat application, thanks to its built-in infrastructure, including even an HTTP server. However, for our backend needs, we require something more powerful in functionality. That's where the Hono library steps in, offering a minimalist and flexible approach to web application development.

Introducing Hono

Hono is a modern, lightweight web framework designed for JavaScript and TypeScript developers. It's built to give you just what you need for web development without providing more features than necessary and taking choices away from you.

Hono handles essential tasks such as routing, data validation, and middleware creation but doesn't enforce other things, such as database interactions, architectural patterns, or logging libraries. This flexibility lets you choose the best tools and practices for your project where you need to, but also provides a good number of tools to do the boring things.

The main features of Hono are as follows:

- **TypeScript integration**: Hono integrates smoothly with TypeScript, providing additional type safety around your endpoints

- **Routing**: It provides straightforward route handling, which helps us to identify which endpoint to call based on the URL

- **Middlewares**: Hono supports middleware that lets you add functionality before and after the request handler executes

- **Error handling**: The library includes built-in error handling to streamline debugging and provide meaningful errors to the caller

- **JSON support**: Hono has native support for JSON parsing and response formatting is a part of Hono, so we can easily accept and return the most popular API format

- **Query and parameter parsing**: It automatically parses URL parameters and query strings

- **Static file serving**: Hono can serve static files, such as images, CSS, and JavaScript

- **Customizable context**: You can extend the request context with custom properties or methods

- **Cookie handling**: The framework provides functions for setting, getting, and deleting cookies, which are often used for authentication purposes

- **Headers manipulation**: Hono simplifies the manipulation of HTTP request and response headers, which is important for security

- **Additional libraries**: Hono complements its core functions with external libraries, mainly as middleware, for functionalities such as CORS setup, request logging, and tracking endpoint performance metrics

As we develop our app, we will get more familiar with Hono and Bun and what they can do, so let's set up our project and see them in action.

Setting up the project

Let's begin by setting up the core part of our project. We'll make this easier by using templates, which help us quickly create the basic structure:

1. Start by running this command to create a Hono project from a template using Bun:

   ```
   $ bun create hono chat_backend
   ```

2. When prompted, select the bun template and hit *Enter*. In a few seconds, you'll have a minimal setup ready. Next, navigate to the project folder, install the necessary dependencies, and launch the local backend server with these commands:

   ```
   $ cd chat_backend
   $ bun install
   $ bun run dev
   ```

3. After this, you can expect the following output in our terminal:

   ```
   $ bun run --hot src/index.ts
   Started server http://localhost:3000
   ```

Now, if you visit http://localhost:3000 in your browser, you should see the message **Hello Hono!** displayed. Congratulations, you've just created your first Hono application! It's pretty basic for now, but we'll be adding more to it shortly.

Let's take a look at the `package.json` file to understand its contents:

package.json

```
{
  "scripts": {
    "dev": "bun run --hot src/index.ts"
  },
  "dependencies": {
    "hono": "^3.12.2"
  },
  "devDependencies": {
    "@types/bun": "^1.0.0"
  }
}
```

In this file, we can see a `scripts` key, which defines what command we will be able to run. The `dev` script is what Bun runs. It's straightforward – it starts our project using the entry point `src/index.ts`. A notable feature here is the `--hot` flag, which enables **hot reloading**. This means that Bun automatically reloads files that have been changed without restarting the entire OS process. This makes for a much faster and more pleasant development experience.

Regarding dependencies, we currently have one production dependency, `hono`, and one development dependency, `@types/bun`. The `@types/bun` package is a TypeScript pattern that you'll often see. It provides type definitions that are specific to the Bun environment, such as the return types of Bun's functions.

Now, let's check out the TypeScript configuration in `tsconfig.json`. We'll remove JSX-related configurations (as we do not use JSX in the backend) and add Bun types, leading to the following content:

tsconfig.json

```
{
  "compilerOptions": {
    "strict": true,
    "esModuleInterop": true,
    "types": ["bun-types"]
  }
}
```

Let's break down what each part of `compilerOptions` does:

- `"esModuleInterop": true`: This setting allows for the compatible use of CommonJS modules in TypeScript, similar to ES6 modules. It simplifies importing CommonJS modules, enabling syntax such as `import fs from 'fs'` instead of `import * as fs from 'fs'`.

- `"strict": true`: Activates all strict type-checking options in TypeScript. This leads to the most thorough type checking, including strict `null` checks and no implicit `any`, among others. It's a way to ensure more comprehensive error checking during code compilation.

- `"types": ["bun-types"]`: Specifically includes the Bun type library for the compiler. Usually, you might not need to specify types like this, but Bun, being a unique environment, requires its types to be explicitly declared in this configuration.

Next, let's discuss the `bun.lockb` file. This file is a lockfile utilized in Bun. Its main role is to maintain consistency in your project's dependencies by locking down specific versions of each package, along with their transitive dependencies.

For instance, if your `package.json` indicates hono as `"^3.12.2"` (meaning any version from 3.12.2 up to, but not including, 4.0.0), `bun.lockb` might specifically lock it to `"3.13.0"`. This ensures that every developer working on the project is using the exact same version, preventing the common issue of "it works on my computer."

Notably, the `bun.lockb` file is in binary format, which Bun uses to reduce file size and optimize the performance of its package management system.

Now, let's take a look at the final important file, `src/index.ts`, which contains our server's code:

src/index.ts

```
import { Hono } from 'hono'

const app = new Hono()

app.get('/', (c) => {
  return c.text('Hello Hono!')
})

export default app
```

Now, let's break down what's happening in this code:

- **Importing Hono and server initialization**: We start by importing the Hono library. After that, we create a new instance of the Hono server with `const app = new Hono()`.

- **Route definition**: The line `app.get('/',` is where we define a route. It instructs the Hono app to handle `HTTP GET` requests at the root URL path `'/'`. When this URL is accessed, the callback function `(c) => { return c.text('Hello Hono!') }` is executed. This function utilizes the context object `c` provided by Hono to send a text response `Hello Hono!` back to the client.

- **Exporting the app**: `export default app` allows the app instance to be imported and utilized in other files, or it can be used to start the server. However, working with plain text responses isn't very common, so we'll change our endpoint to return a JSON response instead. So, we'll replace `return c.text('Hello Hono!')` with `return c.json({'message':'Hello Hono!'})`.

After making this change and reloading the page, your browser should display the following:

```
{
    "message": "Hello Hono!"
}
```

And here we are, with our first JSON-returning endpoint! Before moving forward with the development of our application, it's a good idea to integrate some tools that will simplify our workflow and enhance our code's quality. We will set up linting, formatting, and utility middleware, and we will also discuss handling environment variables. Let's begin with linting and formatting.

Adding linting and formatting

Linting is the process of scanning code to find errors or inconsistencies without actually running it. It's incredibly useful for catching unused variables, preventing `console.log` statements in production, and even identifying logical errors such as infinite loops.

As with static typing, it's all about making our code better and catching issues early. For this project, we'll use **ESLint**, the most common linter for JavaScript. We'll keep things simple and stick to the rules recommended by the Hono project.

First, let's install `eslint` globally so that we can use it from the command line:

```
$ bun install -g eslint
```

Now let's install the rules recommended by Hono:

```
$ bun install --dev @hono/eslint-config
```

Now, we can create a file in the root of our project with `eslint` configs and put the following configuration into it to include the recommended linting rules by Hono:

.eslintrc.cjs

```
module.exports = {
  extends: ["@hono/eslint-config"]
};
```

This sets you up with the basic `eslint` configuration. We can run our linter with the following command:

```
$ eslint --fix
```

It's beneficial to integrate the linter with your IDE. This allows it to run automatically when you save a file to make the developing experience even smoother.

Here's how you can do it in WebStorm:

1. Open **Settings**.
2. Go to **Languages & Frameworks | JavaScript | Code Quality Tools | ESlint**.
3. Turn on **Automatic ESlint config** if it was turned off.
4. Check the mark on **run eslint --fix on save**.

To enable it in Visual Code, you can do the following:

1. Open **Settings**.
2. Type `editor.codeActionsOnSave` into the search bar at the top.
3. Choose **Edit in settings.json** to open the `settings.json` file in VSCode.
4. Insert these lines into your `settings.json` file:

    ```
    "editor.codeActionsOnSave": { "source.fixAll.eslint": true},
    "eslint.alwaysShowStatus": true
    ```

To see how it works in action, you can create a variable in `src/index.ts` that is not used. Now, save the file and you will see a warning in our IDE that the variable is not used. Also, now you can run `eslint` in your terminal and you are going to get a message similar to this:

```
 8:7  warning  'a' is assigned a value but never used  @typescript-
eslint/no-unused-vars
 ✗ 1 problem (0 errors, 1 warning)
```

Now, let's focus on formatting. **Formatting** ensures that our code has a consistent style, which is incredibly valuable when multiple people are collaborating on a project. It's interesting to note that there are many effective code styles out there. The key isn't which style you choose, but rather that everyone sticks to the same style consistently. A formatting tool that automatically enforces a style is a lifesaver, saving countless hours during code reviews and preventing disputes over code style. For our project, we'll integrate **Prettier**, a popular tool that automatically formats code to a predefined style.

It's also very comfortable to integrate Prettier in `eslint` so that they run together and fix both formatting and linting errors at the same time. This allows both tools to run together, fixing formatting and linting errors simultaneously. First, we will need to install additional libraries to handle Prettier as `eslint` plugins:

```
$ bun install --dev eslint-config-prettier eslint-plugin-prettier
```

And now you can expand our `eslint` configuration to look like this:

.eslintrc.cjs

```
module.exports = {
  extends: ["@hono/eslint-config", "plugin:prettier/recommended",
"prettier"],
  plugins: ["prettier"],
  rules: {
    "prettier/prettier": "error",
  }
};
```

This addition includes `Prettier` rules in the `eslint` configuration. We've set `Prettier` errors to be treated as errors rather than warnings. This strict approach helps maintain the overall quality of the project.

Now, try adding extra tabs or spaces to a variable in your code. When you save the file, you'll notice it automatically formats to the correct style, thanks to `Prettier`. Handy, right?

With linting and formatting set up, let's move on to discussing additional middleware that can further enhance our development experience.

Adding middleware

Middleware in Hono acts as interceptor functions that process the incoming HTTP request before it reaches the final route handler. Middleware can also process responses before they are sent back to the client. They can modify requests (e.g., parse the body, add headers) and responses (e.g., set cookies, modify headers).

Middleware executes in the order it is defined in the code. Each middleware function can decide whether to pass a request to the next piece of middleware or to end the response cycle. It is commonly used for logging, authentication, error handling, and data parsing.

We are going to implement our own middleware to handle authentication. However, for now, let's integrate some ready-made middleware provided by Hono. We'll add one for basic request logging and another to append performance metrics to responses.

Here's how we modify the beginning of our `index.ts` file to include these:

src/index.ts

```
import { Hono } from 'hono'
import { logger } from "hono/logger";
import { timing } from "hono/timing";

const app = new Hono()
app.use("*", timing());
app.use("*", logger());
```

In this code, we import the `logger` and `timing` libraries and then add them to our `app` using the `app.use` function with the path `'*'`. This path specification ensures the middleware is applied to all incoming requests, though it can be adjusted to target specific endpoints.

Now, when you hit the endpoint, the logs will include additional lines such as this:

```
<-- GET /
--> GET / 200 4ms
```

These log entries provide details such as the accessed endpoint, the response status code, and the time taken for the response. Additionally, if you check the response headers in your browser's developer console, you'll notice a new `'Server-Timing'` header attached. It carries similar information, such as this:

```
total;dur=0.1;desc="Total Response Time"
```

This addition offers valuable insights into the performance of our requests.

Now, let's discuss how to manage environment variables effectively.

Handling environment variables

Environment variables are a standard way of keeping secrets or platform-specific details outside of the code base. They're also handy for configuring various aspects of an application. Bun has a built-in mechanism for handling environment variables, which can be accessed in your code through `Bun.env.VARIABLE_NAME`. It even automatically reads from a `.env` file, so all we need to do is provide this file.

Here's how to set it up:

1. First, create a file named `.env`.

2. Inside the `.env` file, define a variable such as `TEST=test value`.

3. In our `src/index.ts`, add the line `console.log(Bun.env.TEST)`. Remember, environment variables aren't hot-reloaded, so you'll need to restart the server to see the changes in your console. After restarting, you should see `test value` printed in the terminal.

Sometimes, you might need different environment files for various scenarios, such as local development versus production. Or, you might have different sets of environment variables for different use cases, so you end up with multiple `.env` files, while Bun will handle only `.env` for you. In such situations, you can use a tool such as `dotenv` to load additional environment variables from multiple files.

Let's illustrate it by logging another environment variable from `.env.dev`. Add `console.log(Bun.env.AI)` to `src/index.ts` and add `AI=chat` in our `.env.dev` file. You will see that the value is undefined. Let's fix this by restarting our server and providing it with additional environment variables files:

```
$ bunx dotenv -e .env -e .env.dev -- bun run dev
```

After restarting the project with this command, you should see both `test value` and `chat` in your console.

With environment variables set up, we're now ready to dive into discussing the project structure for our application.

> **Important note**
> Do not add environment files to your Git repo because it is a security risk. Your environment files must be put into `.gitignore` file, and should not be committed.

Discussing the project structure

When structuring code for a project, it's essential to have a clear separation of concerns. This means organizing files and folders in a way that each component has a distinct responsibility, contributing to overall project clarity and maintainability. For our relatively small project, we'll opt for a simple structure.

Here's the structure we'll use and the rationale behind it:

- `src/controllers`: This folder will contain specific REST endpoint handlers. Each controller deals with incoming requests and generates appropriate responses. By isolating endpoint logic in controllers, we make it easier to update or extend API functionalities.

- `src/middlewares`: Here, we'll store additional middleware functions for Hono. Middleware is crucial for processing requests and responses, offering functionality such as authentication, logging, or data parsing. Keeping them in a dedicated folder allows for easy reuse and management.

- `src/models`: This directory is designated for type definitions of the objects used in our code. It ensures a centralized location for data structure definitions, enhancing code consistency and reducing the likelihood of type-related errors.

- `src/storage`: A folder to manage code that interacts with various storage solutions, such as in-memory databases, SQL databases, or ORMs. This separation ensures that changes in storage logic don't impact other parts of the application.

- `src/constants.ts`: This is a file to hold project-wide constants. By centralizing constants, we ensure uniformity and prevent discrepancies that can arise from hardcoding values in multiple locations.

- `src/index.ts`: This is the entry point of our application. This is where we tie together various components of our application, setting up the server, middleware, routes, and any initial configurations.

- `tests`: This is a dedicated folder for storing test files.

This structure is tailored for our chat backend application, ensuring each module has a clear role and responsibility. It fosters an organized development environment, making it easier to navigate, maintain, and scale our application as needed.

Summary

In this chapter, we explored the essentials of Bun, a rising star in the JavaScript runtime landscape, and Hono, a framework known for its simplicity and effectiveness in web application development.

We walked through the practical steps of setting up a development environment tailored for Bun and Hono. This included the installation process, establishing a coherent code structure, and integrating essential tools such as logging and ESLint. The chapter also covered key topics such as project setup, linting and formatting, middleware integration, managing environment variables, and discussing effective project structure strategies.

In the next chapter, we're going to implement the backend functionality for our chat app. Our backend will use in-memory storage, and we'll have the functionality from start to finish.

4

Building Backend Infrastructure with Bun, Hono, and TypeScript

Having covered the basic building blocks of Hono, Bun, and TypeScript required to build our backend API server, we are now ready to actually implement all the backend functionality required for our chat application.

In this chapter, we will work on building data models, storage, middleware, routes, authentication and authorization mechanisms, and chat and message controllers. All of these things in the context of a chat application provide us with the core knowledge required to build a real-world backend application.

Here are the main topics that we'll explore in detail:

- Implementing in-memory storage
- Implementing authentication and authorization
- Implementing chat controllers

We'll kick off by outlining the data models for our application and putting together an in-memory storage solution.

Technical requirements

For this chapter, we won't need to install any additional libraries. All the code examples we discuss are available in the GitHub repository: `https://github.com/PacktPublishing/Full-Stack-Web-Development-with-TypeScript-5/tree/main/Chapter04`.

Implementing in-memory storage

To develop a working application, we will need to store our data somewhere. In this section, we will define the interfaces we are going to use for our storage and create an in-memory implementation of our storage for users, chats, and messages. In-memory storage is a method of storing data directly in the main memory (RAM) of a computer. Because data stored in memory is volatile, it's lost when the application stops.

Defining the interfaces

Let's start this process by defining the data interfaces that we are going to use in our database classes. In the following code, we will create the structure of all the objects that we are going to use when we interact with the database:

src/models/db.ts

```
export type Email = `${string}@${string}.${string}`;
```

This line defines a custom type named `Email`. It uses template literal types to ensure that any string matching this type must follow the conventional email format: a string, followed by an @ symbol, another string, a dot (.), and finally, another string. This is a handy and powerful way to enforce a basic structure for email addresses.

Next, we define the data interfaces:

```
export interface DBEntity {
  id: string;
  createdAt: Date;
  updatedAt: Date;
}

export interface DBUser extends DBEntity {
  name: string;
  email: Email;
  password: string;
}

export interface DBChat extends DBEntity {
  ownerId: DBUser["id"];
  name: string;
}
```

Let's break down the interfaces:

- DBEntity: This interface acts as a base for other database entity interfaces. It includes three properties: id (a string that uniquely identifies the entity), createdAt (a Date object representing when the entity was created), and updatedAt (a Date object representing when the entity was last updated).

- DBUser: This interface extends DBEntity, meaning it inherits all properties of DBEntity (id, createdAt, and updatedAt) and adds three more: name (the user's name), email (the user's email address, which must match the Email type), and password (the user's password).

- DBChat: This interface is similar to DBUser, in that it extends DBEntity. It represents a chat entity with two additional properties: ownerId (a reference to the ID of the user who owns this chat) and name (the name of the chat).

MessageType defines a union type named MessageType that can either be "system" or "user". This is used to categorize the origin of the messages in the chat:

```
export type MessageType = "system" | "user";
```

DBMessage extends DBEntity and represents a message in a chat. It includes chatId (linking the message to a chat), type (the category of the message, as defined by MessageType), and message (the actual text of the message):

```
export interface DBMessage extends DBEntity {
   chatId: DBChat["id"];
   type: MessageType;
   message: string;
}
```

The following interfaces are used for creating the respective database types. They use TypeScript's Pick utility type to create types that only include a subset of properties from the original interfaces:

```
export type DBCreateUser = Pick<DBUser, "email" | "password" |
"name">;
export type DBCreateChat = Pick<DBChat, "name" | "ownerId">;
export type DBCreateMessage = Pick<DBMessage, "chatId" | "message" |
"type">;
```

For instance, DBCreateUser includes only email, password, and name from DBUser, focusing on the properties needed when creating a new user, as others will be generated by the code.

The interfaces defined in this section are the interfaces we are going to use when we interact with the database. However, when we operate on the API level, it's good practice to introduce an additional layer of types that we are going to receive and return from our endpoints. In a big application, database types and API types can overlap only partially because both layers handle the same data but with different details and structures suited to their specific roles in the application; for example, we can create multiple database objects from one endpoint. In our app, however, they mostly overlap as we have straightforward **Create, Read, Edit, Delete** (CRUD) endpoints. So, let's create the API types, which we are going to use as a form of input and output data from our endpoints.

Creating database and API types

All of the types we created are simple aliases to our database classes. The only class that is different is `APIUser`, which omits `password` from its definition as it's not something we want to expose when we return a user:

src/models/api.ts

```
import type {
  DBChat,
  DBCreateChat,
  DBCreateMessage,
  DBCreateUser,
  DBMessage,
  DBUser,
} from "./db";

export type APICreateUser = DBCreateUser;
export type APIUser = Omit<DBUser, "password">;

export type APICreateChat = DBCreateChat;
export type APIChat = DBChat;

export type ApiCreateMessage = DBCreateMessage;
export type ApiMessage = DBMessage;
```

With this done, we can move on to implementing our in-memory storage. The storage class we are going to implement will be an abstract class so it can be used as a storage for user information, chats, and messages interchangeably. We will start with defining the `IDatabaseResource` interface, which will represent an abstract way to store and operate data.

Then, we will pass the storage class interface around in the app instead of a concrete implement. It will make it possible to easily replace the specific implementation with in-memory, SQL, or ORM implementation without the need to change the types and our application. This shows the use of abstraction, a practice we introduced in *Chapter 2*:

src/storage/types.ts

```
export interface IDatabaseResource<T, S> {
  create(data: S): Promise<T>;
  update(id: string, data: Partial<S>): Promise<T | null>;
  get(id: string): Promise<T | null>;
  find(data: Partial<T>): Promise<T | null>;
  findAll(data: Partial<T>): Promise<T[]>;
  delete(id: string): Promise<T | null>;
}
```

Here is a definition for a CRUD implementation of a database resource with all the standard methods for data manipulation. It is a generic interface with two generics, T and S, which are generic parameters and placeholders for specific types. It's an example of generics as we discussed in *Chapter 2*, and it will add reusability and flexibility to our code:

- S represents a data type that contains all the fields required to create an entity

- T is the type with all the fields of a database entity

For example, in the create and update methods, we use S as our input type as we only need the fields for creation. But in find, we retrieve an object with any fields from the whole resource. This is because we may search the objects by fields such as id, which we don't provide during the creation or update of the object.

Let's break the other parts of the class down by the idea behind each method:

- create(data: S): Promise<T>: This creates a new entity with input data of the S type, returning a promise that resolves to the created entity of the T type. It uses a Promise type because the method is going to be asynchronous, meaning that it returns Promise with some types.

- update(id: string, data: Partial<S>): Promise<T | null>: This updates an entity identified by id with the provided data. It returns a promise resolving to the updated entity or null. The Partial type here makes it possible to send not the whole object for the update but only some selected fields from the original type.

- get(id: string): Promise<T | null>: This retrieves an entity by its id, returning a promise that resolves to the entity or null.

- find(data: Partial<T>): Promise<T | null>: This finds an entity matching the partial data criteria, returning a promise that resolves to the entity or null.

- `findAll(data: Partial<T>): Promise<T[]>`: This finds all entities matching the partial data criteria, returning a promise that resolves to an array of entities.
- `delete(id: string): Promise<T | null>`: This deletes an entity by its `id`, returning a promise that resolves to the deleted entity or `null`.

You also see that we wrap our return types in the `Promise` type. This is the way to work with async functions in TypeScript; if our function is asynchronous, its return type will always be `Promise<T>`.

In-memory implementation of the interfaces

Now, we can turn to the concrete in-memory implementation of the `IDatabaseResource` interface.

The `T extends S & DBEntity` generic type constraint means that `T` must be a type that extends both `S` and `DBEntity`:

storage/inmemory.ts

```
import type { DBEntity } from "../models/db";
import type { IDatabaseResource } from "./types";

export class SimpleInMemoryResource<T extends S & DBEntity, S>
  implements IDatabaseResource<T, S>
{
```

So, our entity type must include all the fields from the basic `DBEntity` and the type used for its creation. An example of `T` can be `DBChat` and an example of `S` can be `DBCreateChat`. Such a structure will make sure that we use the correct types for the creation and retrieval of the data.

We use an array data of the `Array<T>` type to store entities in memory. This simulates a database table:

```
data: Array<T> = [];
```

Here is the `create` method:

```
async create(data: S): Promise<T> {
  const fullData = {
    ...data,
    id: this.data.length.toString(),
    createdAt: Date.now(),
    updatedAt: Date.now(),
  } as T;
  this.data.push(fullData);
  return fullData;
}
```

Let's discuss what the `create` method does:

- It generates a unique `id` using the array's length, ensuring that each entity has a distinct identifier. The ID is generated as the size of the array, and this approach will result in the duplication of the ID if we delete an object before adding a new one. It will do for now as in-memory implementation is not used in production.

- It assigns `createdAt` and `updatedAt` using `Date.now()`, providing timestamps.

- It spreads data into `fullData` to include all necessary fields, casting it as the T type.

- It adds `fullData` to the data array, simulating data insertion.

Here is the `delete` method:

```
async delete(id: string): Promise<T | null> {
  const entity = this.data.find((x) => x.id === id);
  if (entity) {
    this.data = [...this.data.filter((x) => x.id !== entity.id)];
    return entity;
  } else {
    return null;
  }
}
```

The `delete` method locates an entity by `id` using `Array.find` – if found, it filters out the entity from the data, effectively removing it. If not found, it returns the deleted entity or `null`.

Next, we have the `get` method:

```
async get(id: string): Promise<T | null> {
  return this.data.find((x) => x.id === id) || null;
}
```

The `get` method retrieves an entity by `id` using `Array.find` – it returns the entity or returns `null` if the entity doesn't exist.

Now, let's finish up the methods in the class.

Here is the `find` method:

```
async find(data: Partial<T>): Promise<T | null> {
  return (
    this.data.find((x) => {
      for (const key in data) {
        if (data[key] != x[key]) return false;
      }
      return true;
```

```
      }) || null
    );
  }
```

The `find` method searches for entities matching data (partial `T` type), and returns the first match or `null`.

`findAll` returns an array of all matches. It uses a loop to compare each key in data with entities in the `data` array:

```
async findAll(data: Partial<T>): Promise<T[]> {
  const res = this.data.filter((x) => {
    for (const key in data) {
      if (data[key] != x[key]) return false;
    }
    return true;
  });
  return res;
}
```

The `update` method retrieves the existing entity by `id`:

```
async update(id: string, data: S): Promise<T | null> {
  const entity = await this.get(id);
  if (entity) {
    const newEntity = { ...entity, ...data, updatedAt: Date.now() };
    await this.delete(id);
    this.data.push(newEntity);
    return newEntity;
  } else {
    return null;
  }
}
}
```

The `update` method merges the existing entity with new data and updates the `updatedAt` timestamp. It deletes the old entity, pushes the updated entity into data, and returns the updated entity or `null` if not found.

With this, we've finished the interface for our storage and an in-memory implementation we are going to use in our app. Let's now turn to implementing authentication and authorization.

Implementing authentication and authorization

Our app requires certain security measures: we need to block some endpoints from users who aren't logged in, and we also need to know who the user is and what permissions they have when they're using a controller. We'll handle this by setting up **JSON Web Token (JWT)** authentication, based on Hono's JWT authentication module.

Here's how our app will work:

- Users can access both login and register endpoints unauthenticated
- For any other endpoint, users must have a valid JWT token in the Authorization Headers
- When there is a valid token present, we will attach the user's ID to the request context, so that we can easily retrieve it in our controllers

To register, users must provide an email, password, and name. The email cannot be already associated with another user. In line with best security practices, passwords will not be stored directly. Instead, we'll store only the hash of the password.

With this plan in mind, we'll start by creating the **authentication** middleware.

Developing the authentication middleware

First off, we'll set the secret for our JWT token in our `.env` file:

.env

```
JWT_SECRET=YOUR_SECRET_VALUE_32_CHARS_LONG
```

`JWT_SECRET` is used to encrypt and decrypt the signature of our JWT. We can now access `JWT_SECRET` during code execution using either Bun or Hono's built-in methods. We'll use Hono's method, as it works across different environments besides Bun, such as Cloudflare Workers. With this done, we can move forward with discussing and creating authentication middleware.

In Hono, **middleware** consists of functions that take a `Context` instance and the `next` function as arguments. The `next` function, when called, hands over the control to the next middleware in the chain or the route handler if it's the last middleware. Middleware can perform actions before and after calling `next`, allowing for both the preprocessing and post-processing of requests. With this knowledge, we can write our middleware code. We will first discuss the constants that we will need to declare and then we will continue with the middleware:

src/constants.ts

```
export const API_PREFIX = "/api/v1";

export type ContextVariables = { Variables: { userId: string } };
```

`API_PREFIX` will be used as the URL prefix for all our URLs. As you see, we add versions to our endpoints so that we can maintain multiple versions at the same time in the future.

`ContextVariables` defines what kind of variables we will attach to our endpoint request object. We will see it in action soon in the *Defining our controllers* section.

Now, we can proceed with our middleware implementation. We will define two middleware functions: `checkJWTAuth` and `attachUserId`.

First, we import a few things:

src/middlewares/auth.ts

```
import type { Context } from "hono";
import { env } from "hono/adapter";
import { jwt } from "hono/jwt";
import { API_PREFIX } from "../constants";
import { AUTH_PREFIX, LOGIN_ROUTE, REGISTER_ROUTE } from "../
controllers/auth";

import type { APIUser } from "../models/api";
```

Here are the important terms seen in the preceding code block:

- `Context`: Imported from Hono, it represents the context of a single request or response cycle.

- `env`: A utility from `hono/adapter` to access environment variables.

- `jwt`: Middleware from `hono/jwt` for JWT authentication.

- `AUTH_PREFIX`: This is the prefix we are going to use for authentication routes. This and the following prefixes are going to be defined in the `../controllers/auth` file that we will discuss later in this section.

- `LOGIN_ROUTE` and `REGISTER_ROUTE`: These are the specific route URLs for our login and register endpoints, respectively.

Then, we implement the checkJWTAuth middleware. We use it to enforce JWT-based authentication for all routes except the login and register routes:

```
export async function checkJWTAuth(
  c: Context,
  next: () => Promise<void>,
): Promise<Response | void> {
  if (
    c.req.path === API_PREFIX + AUTH_PREFIX + LOGIN_ROUTE ||
    c.req.path === API_PREFIX + AUTH_PREFIX + REGISTER_ROUTE
  ) {
    return await next();
  } else {
    const { JWT_SECRET } = env<{ JWT_SECRET: string }>(c);
    const jwtMiddleware = jwt({
      secret: JWT_SECRET,
    });
    return jwtMiddleware(c, next);
  }
}
```

checkJWTAuth first checks whether the current request's path matches either the login or register route. This is done by concatenating API_PREFIX, AUTH_PREFIX, and the specific route constants. If the request is for login or register, it bypasses JWT validation (next() is called without any JWT check).

For all other routes, checkJWTAuth retrieves JWT_SECRET from the environment variables and initializes the JWT middleware with this secret. The JWT middleware is then invoked with the current Context and the next function. This middleware validates the JWT from the request, ensuring secure access to protected routes.

attachUserId is a middleware function that we use to extract the user's ID from the JWT payload and attach it to the context:

```
export async function attachUserId(
  c: Context,
  next: () => Promise<void>,
): Promise<Response | void> {
  const payload = c.get("jwtPayload") as APIUser;
  if (payload) {
    const id = payload.id;
    c.set("userId", id);
  }
  await next();
}
```

`attachUserId` retrieves the JWT payload from the context, which was decoded by the JWT middleware. If the payload is present (indicating a valid JWT), it extracts the user's ID from this payload. This ID is then attached to the context (`c.set("userId", id)`) for use in subsequent middleware or route handlers.

Defining our controllers

Let's proceed with defining our controllers with endpoints to log in and register.

In this piece of code, we will define the two routes that will accept and use the needed data for registration and logging in:

src/controllers/auth.ts

```
import { Hono } from "hono";
import { env } from "hono/adapter";
import { sign } from "hono/jwt";
import type { ContextVariables } from "../constants";
import type { DBCreateUser, DBUser } from "../models/db";
import type { IDatabaseResource } from "../storage/types";
```

`sign` is the JWT signing function from Hono's JWT utilities. This function is used to create a JWT.

This is the prefix for our application here:

```
export const AUTH_PREFIX = "/auth/";
```

We use `ContextVariables`, which we defined in the *Developing the authentication middleware* section, to signal that we will attach `userId` to our requests; it will add correct types when we retrieve `userId` for the compiler:

```
export const authApp = new Hono<ContextVariables>();
```

These are the prefixes for our authentication endpoint:

```
export const LOGIN_ROUTE = "login/";
export const REGISTER_ROUTE = "register/";
```

Here, we define a constant string that we are going to return from our endpoints in case of errors:

```
export const ERROR_USER_ALREADY_EXIST = "USER_ALREADY_EXIST";
export const ERROR_INVALID_CREDENTIALS = "INVALID_CREDENTIALS";
```

Here, we accept a `userResource` object that implements an `IDatabaseResource` interface of the user type:

```
export function createAuthApp(
  userResource: IDatabaseResource<DBUser, DBCreateUser>,
) {
```

We pass `DbUser` and `DbCreateUser` as the specific types for the `T` and `S` types we use in `IDatabaseResource`. This is where we will pass our in-memory storage implementation.

Now, let's continue with implementing the authentication endpoints. Here, we have the registration endpoint:

```
authApp.post(REGISTER_ROUTE, async (c) => {
  const { email, password, name } = await c.req.json();
  if (await userResource.find({ email })) {
    return c.json({ error: ERROR_USER_ALREADY_EXIST }, 400);
  }
  const hashedPassword = await Bun.password.hash(password, {
    algorithm: "bcrypt",
  });
  await userResource.create({ name, email, password: hashedPassword
});
  return c.json({ success: true });
});
```

First, we extract user data (`email`, `password`, and `name`) from the request body. Then, we check whether a user with the given email already exists using `userResource.find`. If the user exists, we return an error response. Otherwise, we hash the password using `Bun.password.hash` (a very useful function from Bun with cutting-edge `Argon2Id` algorithms to hash our password, which provides additional resistance from cracking the password using GPUs and special hardware), and create a new user using `userResource.create`.

Next, we have the login endpoint:

```
authApp.post(LOGIN_ROUTE, async (c) => {
  const { email, password } = await c.req.json();
  const fulluser = await userResource.find({ email });
  if (
    !fulluser ||
    !(await Bun.password.verify(password, fulluser.password))
  ) {
    return c.json({ error: ERROR_INVALID_CREDENTIALS }, 401);
  }
```

```
      const { JWT_SECRET } = env<{ JWT_SECRET: string }, typeof c>(c);
      const token = await sign({ ...fulluser, password: undefined },
JWT_SECRET);
      return c.json({ token });
    });
    return authApp;
}
```

First, we retrieve `email` and `password` from the request body. Then, we find the user by email. If not found or if the password verification (using `Bun.password.verify`, which hashes the incoming value and compares it with the existing hash) fails, we return an error response. If authentication is successful, we generate a JWT token using `sign`, omitting the password from the token payload. Finally, we send back a response with the JWT token.

With authentication in place, we can introduce other controllers and, finally, instantiate our server.

Implementing chat controllers

Now, we have come to the implementation of the main functionality for our application that we are going to use to provide the core features of our app with the help of controllers.

For our app, we will need to support the following functionality:

- **Create a new chat**: Users can start a new chat on some topic in our app
- **Get all chats**: Users can see all the created chats in a list on the menu
- **Get a specific chat**: Users can enter a specific chat and the details
- **List of chat messages**: Users can retrieve all the messages in a chat
- **Create a new message in a chat**: Users can send new messages in a chat

Let's implement all this functionality in our endpoints.

Implementing the endpoints

In this piece of code, we will create our chat app that is going to have endpoints for chat and message creation as well as their retrieval:

src/controllers/chat.ts

```
import { Hono } from "hono";
import type { ContextVariables } from "../constants";
import type {
  DBChat,
```

```
  DBCreateChat,
  DBCreateMessage,
  DBMessage,
} from "../models/db";
import type { IDatabaseResource } from "../storage/types";
```

Here are the prefixes we are going to use in our URLs:

```
export const CHAT_PREFIX = "/chat/";
const CHAT_ROUTE = "";
const CHAT_MESSAGE_ROUTE = ":id/message/";
```

`CHAT_PREFIX`, `CHAT_ROUTE`, and `CHAT_MESSAGE_MESSAGE` and HTTP methods, such as `GET` and `POST`, give us the `REST` style of our endpoints. Simply put, our URL should identify the resource it works on, and HTTP methods must describe the action that is performed on the endpoint. Let's give the basic examples of the possible canonical routes for CRUD operations:

- `GET /chat/`: Get a list of chats

- `POST /chat/`: Create a chat

- `GET /chat/:id`: Get one chat

- `POST` or `PUT /chat/:id`: Update one chat

- `DELETE /chat/:id`: Delete one chat

Now, we can write a function to create our chat app:

```
export function createChatApp(
  chatResource: IDatabaseResource<DBChat, DBCreateChat>,
  messageResource: IDatabaseResource<DBMessage, DBCreateMessage>,
) {
  const chatApp = new Hono<ContextVariables>();
```

The `REST` endpoints function receives two resources: `chatResource` and `messageResource`. These are two in-memory resources for chats and messages, respectively.

Now, let's finish the implementation of the endpoints:

```
chatApp.post(CHAT_ROUTE, async (c) => {
  const userId = c.get("userId");
  const { name } = await c.req.json();
  const data = await chatResource.create({ name, ownerId: userId });
  return c.json({ data });
});
```

This endpoint allows users to create new chats. A user's ID (`userId`) is retrieved from the context. The chat's name is obtained from the request's JSON body. The `chatResource.create` method is used to create a new chat, linking it to the user who created it.

The `chatApp.get(CHAT_ROUTE, async (c) => {...})` endpoint lists all chats owned by a user:

```
chatApp.get(CHAT_ROUTE, async (c) => {
  const userId = c.get("userId");
  const data = await chatResource.findAll({ ownerId: userId });
  return c.json({ data });
});
```

`chatApp.get` uses the user's ID to find all chats associated with that user through `chatResource.findAll`.

`chatApp.get(CHAT_MESSAGE_ROUTE, async (c) => {...})` retrieves all messages for a specific chat:

```
chatApp.get(CHAT_MESSAGE_ROUTE, async (c) => {
  const { id: chatId } = c.req.param();
  const data = await messageResource.findAll({ chatId });
  return c.json({ data });
});
```

`chatApp.get` uses the chat room's ID (`chatId`), extracted from the request URL parameters, which is defined as `chatId` in `CHAT_MESSAGE_ROUTE`, to find messages related to that chat room.

`chatApp.post(CHAT_MESSAGE_ROUTE, async (c) => {...})` allows creating new messages to a chat room:

```
chatApp.post(CHAT_MESSAGE_ROUTE, async (c) => {
  const { id: chatId } = c.req.param();
  const { message } = await c.req.json();

  const userMessage: DBCreateMessage = { message, chatId, type:
"user" };
  await messageResource.create(userMessage);

  const responseMessage: DBCreateMessage = {
    message: "dummy response",
    chatId,
    type: "system",
  };
```

```
      const data = await messageResource.create(responseMessage);

      return c.json({ data });
  });
  return chatApp;
}
```

After a user message is posted, `chatApp.post` automatically inserts a *dummy response*. This is a placeholder for our future GPT integration in *Chapter 11*.

Now, we can write code to glue together our chat and authentication endpoints.

Combining the endpoints

In this code, we will first import the required functions and middleware to construct our whole app, and then we are going to introduce a function that creates our main app:

src/controllers/main.ts

```
import { Hono } from "hono";
import { showRoutes } from "hono/dev";
import { logger } from "hono/logger";
import { timing } from "hono/timing";
import type { ContextVariables } from "../constants";
import { API_PREFIX } from "../constants";
import { attachUserId, checkJWTAuth } from "../middlewares/auth";
import type {
  DBChat,
  DBCreateChat,
  DBCreateMessage,
  DBCreateUser,
  DBMessage,
  DBUser,
} from "../models/db";
import { SimpleInMemoryResource } from "../storage/in_memory";
import { AUTH_PREFIX, createAuthApp } from "./auth";
import { CHAT_PREFIX, createChatApp } from "./chat";
```

Here, we define the function that will create the main app, which is going to encompass our authentication and chat apps:

```
export function createMainApp(
  authApp: Hono<ContextVariables>,
```

```
    chatApp: Hono<ContextVariables>,
) {
```

We pass `API_PREFIX` here, which is going to be applied before all endpoint URLs:

```
const app = new Hono<ContextVariables>().basePath(API_PREFIX);
```

Here is middleware to add headers to our request that will record how much time was spent:

```
app.use("*", timing());
```

Here is another middleware that logs incoming requests:

```
app.use("*", logger());
```

These are our custom middleware for handling JWT authentication and attaching the user's ID to the context:

```
app.use("*", checkJWTAuth);
app.use("*", attachUserId);
```

`checkJWTAuth` and `attachUserId` are applied globally, ensuring that protected routes are accessed only by authenticated users and that user-specific data is easily accessible in route handlers.

These lines mount `authApp` and `chatApp` (our sub-applications) on specific route prefixes:

```
app.route(AUTH_PREFIX, authApp);
app.route(CHAT_PREFIX, chatApp);
```

Here is a handy utility in development mode to display all available routes of the application. This can help us understand the route structure and ensure correct configurations:

```
    showRoutes(app);

    return app;
}
```

The `createInMemoryApp()` function creates the main app with in-memory resources for users, chats, and messages, which we will use to create the entry point of our application. It leverages `SimpleInMemoryResource` to simulate database operations:

```
export function createInMemoryApp() {
  return createMainApp(
    createAuthApp(new SimpleInMemoryResource<DBUser, DBCreateUser>()),
    createChatApp(
```

```
      new SimpleInMemoryResource<DBChat, DBCreateChat>(),
      new SimpleInMemoryResource<DBMessage, DBCreateMessage>(),
    ),
  );
}
```

The only missing step is to call the function we defined earlier in this section in our index file. Let's do it.

Using the index file for calling the function

Here, we simply utilize the function that we defined in the previous file, which is going to be the entrypoint of our application:

src/index.ts

```
import { createInMemoryApp } from "./controllers/main";

const app = createInMemoryApp();

export default app;
```

Now, run the following:

```
$ bun run dev
```

You will see all the routes that we have defined:

```
POST   /api/v1/auth/register/
POST   /api/v1/auth/login/
POST   /api/v1/chat/
GET    /api/v1/chat/
GET    /api/v1/chat/:id/message/
POST   /api/v1/chat/:id/message/
```

As well as that, we can interact with our application using `curl`. First, we will register and log in with a new user, then create and retrieve a chat for the user.

> **Important note**
>
> Be careful to put the closing slash to all our URLs, as our backend considers `/register` and `/register/` to be two separate URLs.

In the next command, we will use `curl` to access our endpoint with the JSON parameters required to register a user:

```
$ curl -X POST http://localhost:3000/api/v1/auth/register/ -H
"Content-Type: application/json" -d "{\"email\": \"user@mail.com\",
\"password\": \"pass\", \"name\": \"Test\"}"
```

You will see the following as the response:

```
{"success":true}
```

So, we've successfully registered our user. Now, let's log in:

```
$ curl -X POST http://localhost:3000/api/v1/auth/login/ -H "Content-
Type: application/json" -d "{\"email\": \"user@mail.com\",
\"password\": \"pass\"}"
```

You should see this response in the terminal:

```
{"token":"$JWT_TOKEN"}
```

Now, you need to copy the $JWT_TOKEN value that you got to the header in the next request:

```
$ curl -X POST http://localhost:3000/api/v1/chat/ -H "Content-Type:
application/json" -H "Authorization: Bearer $JWT_TOKEN" -d "{\"name\":
\"Chat1\"}"
```

You will get the newly created chat in response:

```
{"data":{"name":"Chat1","ownerId":"0","id":"0","createdAt":
1706359408285,"updatedAt":1706359408285}}
```

Now, let's make our final API request and retrieve the list of all chats that belong to us:

```
curl -X GET http://localhost:3000/api/v1/chat/ -H "Authorization:
Bearer $JWT_TOKEN"
```

You will the array of all chats as the result:

```
{"data":[{"name":"Chat1","ownerId":"0","id":"0","createdAt":
1706359408285,"updatedAt":1706359408285}]}
```

So, now we have a fully working backend for our chat; congrats on this! I suggest you play more with the endpoint to create different chats and messages and retrieve them so that you can see how it works.

There is a catch though. If we stop and start our application again, you will see that the data is gone when you call the endpoints. Our users no longer exist and neither do chats. This is because we use in-memory storage that only resides in our operating memory. To fix it, we will introduce persistent storage in the following chapters.

Summary

In this chapter, we've implemented the backend part for our chat application and have gained an understanding of the essentials of backend development, such as routing, controllers, middleware, authentication, data models, and storage. This knowledge helps us to develop real-world robust and maintainable backend applications with all the core parts.

The next chapter will add an additional layer of useful backend techniques, as we will focus on validating the incoming data and on testing our application, techniques that help reduce the number of errors during the development and increase the overall reliability of backend applications.

5

Improving Reliability – Testing and Validation

In *Chapter 4*, we put together the endpoints we need for our chat app and got the basics of our backend in place with Hono, Bun, and TypeScript. Now, it's time to make sure everything runs correctly.

First up, we'll use Zod, which helps us to check whether an object corresponds to the schema we want it to be, to validate the data coming into our endpoints. This means checking that the data is in the correct format and follows the rules we set. Then, we'll get to grips with Bun's testing module. Learning how to use Bun's testing module and Zod will give us the tools to make sure our app can be as reliable as it needs to be, and help us catch any issues early.

In this chapter, we are going to cover the following topics:

- Writing validation with Zod
- Developing tests with Bun

We will start by explaining why we actually need validation, what Zod is, and how to integrate and use it to validate incoming data in our endpoints.

Technical requirements

For this chapter, we will need to add the Zod library for validation. We can add it to our project by running the following command in the terminal:

```
$ bun add zod @hono/zod-validator -d
```

All the code examples we discuss are available in the GitHub repository: https://github.com/PacktPublishing/Full-Stack-Web-Development-with-TypeScript-5/tree/main/Chapter05.

Writing validation with Zod

Validating incoming data to our endpoints is crucial for maintaining the integrity and security of our applications. Proper validation acts as a first line of defense, filtering out malformed, corrupt, or malicious data before it can interact with our systems. It ensures that the data aligns with our expectations and requirements, safeguarding the application from unexpected behavior, crashes, or security vulnerabilities.

By rigorously checking incoming data, we not only protect our backend processes and databases but also provide a more reliable and user-friendly experience, especially in cases when our API is used by third parties. Let's turn now to how we can use it in our app.

So, how can we add it to our application? The library we will use is called **Zod**. Zod is a TypeScript-first schema declaration and validation library that enables us to define the shape and constraints of data in a clear, concise manner. It provides a powerful and flexible way to ensure that the data your application handles is correctly structured and adheres to specified rules.

We will use it as a Hono validator, a form of middleware defined directly on the endpoint that helps check the incoming data. `@hono/zod-validator` is Hono middleware that integrates Zod's validation capabilities into our Hono applications. It allows you to validate incoming request data against predefined Zod schemas.

One additional pleasant aspect is that all this works seamlessly with our types, and when we retrieve the data in our endpoint from the request body or query params it will be annotated with the type we define in our Zod schema.

Let's now add validation to our authentication endpoints, so we see how it works in action.

Adding validation to our authentication endpoints

We will first define the schema for the body of registration and login endpoints, and then we will add the schema to the endpoint definition. Eventually, we can retrieve the data from the validated dictionary, `c.req.valid`, appended to our request object. Most of the code is still the same; so I will replace code that we used previously with . . . .

In the following segment of the code, we're integrating Zod and Hono's `@hono/zod-validator` to validate incoming data for our authentication endpoints. First, we will add the `import` statements:

src/controllers/auth.ts

```
. . .
import { zValidator } from "@hono/zod-validator";
import { z } from "zod";
```

Here, we import the Zod library to validate schemas, and then we import zValidator, a middleware for Hono that uses Zod for schema validation.

Next, we will define the schemas:

```
const registerSchema = z.object({
  email: z
    .string()
    .email()
    .transform((x) => x as Email),
  password: z.string().min(1),
  name: z.string().min(1),
});

const loginSchema = z.object({
  email: z
    .string()
    .email()
    .transform((x) => x as Email),
  password: z.string().min(1),
});
```

Let's explain the preceding code block:

- `const registerSchema = z.object({...});` and `const loginSchema = z.object({...});`: These lines define the validation schemas for the registration and login data using Zod. They ensure that the email is in a proper format and that the password and name fields are strings with a minimum length of 1.

- `.transform((x) => x as Email)`: This part casts the validated email string to the Email type, ensuring type safety.

Now, we will use zValidator in Hono routes and get the validated data in our login endpoint:

```
export function createAuthApp(
  userResource: IDatabaseResource<DBUser, DBCreateUser>,
) {
  authApp.post(
    REGISTER_ROUTE,
    zValidator("json", registerSchema),
    async (c) => {
      const { email, password, name } = c.req.valid("json");
      ...
```

In the preceding code block, we have the following:

- `zValidator("json", registerSchema)`: This is used in the Hono route definition for the registration endpoint. It tells Hono to validate the incoming JSON request body against the `registerSchema`. If the validation fails, the request is automatically rejected and an error response is sent.

- `const { email, password, name } = c.req.valid("json");`: After passing through the `zValidator`, the request's JSON body is validated. The `.valid("json")` method extracts the validated data, ensuring that the `email`, `password`, and `name` variables conform to the structure and constraints defined in the Zod schema.

Next, we will also add the validation to our login endpoint:

```
authApp.post(LOGIN_ROUTE, zValidator("json", loginSchema), async (c)
=> {
    const { email, password } = c.req.valid("json");
    ...
  });
  return authApp;
}
```

`zValidator("json", loginSchema)` is used in the login route and validates the request body against the `loginSchema`. Then, we retrieve `email` and `password` from the validated data.

In a similar way, let's add validation to our chat endpoints.

Adding validation to our chat endpoints

Here, we will also add a validation layer to our chat endpoints. We will begin by adding the required imports and defining the Zod schema for endpoint parameters and body:

src/controllers/chat.ts

```
...
import { zValidator } from "@hono/zod-validator";
import { z } from "zod";

const idSchema = z.object({
  id: z.string().min(1),
});

const chatSchema = z.object({
  name: z.string().min(1),
```

```
});

const messageSchema = z.object({
  message: z.string().min(1),
});
```

Here are the schemas shown in the preceding code block:

- idSchema validates that the id parameter exists and is a non-empty string. It's used to validate path parameters in routes where an id is required.
- chatSchema ensures that the chat name is a non-empty string. It's used to validate the JSON body of requests related to creating a chat.
- messageSchema is for validating the message content, ensuring that the message is a non-empty string. It's used in routes where a message is sent.

Now, we will add these schemas to our endpoints.

We will use chatSchema in our chat creation endpoint to validate that the passed body is correct:

```
. . .
  chatApp.post(CHAT_ROUTE, zValidator("json", chatSchema), async (c)
=> {
    const userId = c.get("userId");
    const { name } = c.req.valid("json");
    . . .
  });
```

We can also use the validation for our query parameters:

```
. . .
  chatApp.get(CHAT_DETAIL_ROUTE, zValidator("param", idSchema), async
(c) => {
    const { id } = c.req.valid("param");
    . . .
  }); chatApp.get(CHAT_MESSAGE_ROUTE, zValidator("param", idSchema),
async (c) => {
    const { id: chatId } = c.req.valid("param");
    . . .
  });
```

In our chat and message GET endpoints, we need to validate that the query parameter id that we pass is in the correct form. We also use the param key to retrieve the validated query parameter. We can also combine validation of the query parameters and the body.

In our message creation endpoint, as seen in the following code snippet, we utilize both the validation of the body and the parameters, and we can later access them with the `param` and `json` key respectively:

```
chatApp.post(
  CHAT_MESSAGE_ROUTE,
  zValidator("param", idSchema),
  zValidator("json", messageSchema),
  async (c) => {
    const { id: chatId } = c.req.valid("param");
    const { message } = c.req.valid("json");
    ...
  },
);
return chatApp;
}
```

Now, if we send the requests to the endpoint, but the data we pass doesn't correspond to the schema we defined, we will get a 400 error back with a description of which fields were incorrect. We will see it more when we write the tests for the validation in our endpoints.

And now let's talk about why we need tests, and what Bun's tests bring to us, and then write a few to cover our endpoints.

Developing tests with Bun

In web development, developing tests is not just a good practice; it's a must for creating reliable and robust applications. Tests act like a safety net, catching errors before they make it to production, where they can be costly and damaging. They ensure that each part of your application works as expected and continues to do so, even as you add new features or refactor old ones. Automating tests lets your team concentrate on writing new code rather than correcting avoidable bugs. Furthermore, detailed testing and extensive coverage improve team efficiency by promoting independence and creating trust through the development of dependable code between teams.

In this book, we've taken the approach of developing the tests after the actual implementation, which I think works better from an explanation perspective, but in the production project scenario, I often prefer to write tests before the implementation. This practice, known as **Test-Driven Development (TDD)**, serves a few important purposes:

- It makes you think about the structure of your code before you begin to write it. This prevents you from realizing that it is not going to work after you've written hundreds of lines of code already.

- It makes you write your code in a testable manner, which almost always results in more granular code with clear segregation.

- TDD often works well for pair programming, as one person can write the tests while the other one focuses on the implementation.

But no matter which approach you take, either TDD or writing the tests afterward, they both will help you to produce fewer bugs and more maintainable code.

There are also different levels of testing that we should talk about, which form the cornerstone of software quality assurance:

- **Unit testing** examines individual pieces of code, typically functions or methods, to ensure they work correctly in isolation. Unit tests are small and typically easy to write.

- **Integration testing** goes a step further by testing how different parts of the application, such as modules or services, interact with each other. Integration tests are a little bit bigger and run more slowly than unit tests but typically produce more code coverage per test.

- **End-to-end** (E2E) **testing** evaluates the entire application's flow from beginning to end, ensuring the user experience is as designed. E2E tests are typically even harder to write, but they are the only ones that can assure that the app actually works completely.

In this chapter, we chose to focus on integration testing as a good balance between how hard it is to write a test and how much coverage we will get per test. This said, let's actually see how we can write code with Bun.

Writing tests with Bun's test runner

Bun provides us with a **test runner** out of the box. The test runner aims to provide a zero-configuration experience, integrate seamlessly with code, and be fast. Here are the main functionalities of the Bun test runner and how it compares to the most popular option for testing in a JavaScript environment – **Jest**:

- **Speed**: Buns' test runner is built for performance. It's designed to run tests extremely quickly, using the speed of the Bun runtime. Compared to Jest, which can be heavy and slow, especially on larger projects, Bun's test runner offers a more efficient and time-saving experience.

- **Concurrency**: It supports concurrent test execution out of the box. This means your tests can run in parallel, significantly cutting down the total test execution time. Bun's test runner is optimized to make concurrency straightforward and efficient.

- **Zero configuration**: Bun test runner aims for a zero-configuration experience. You can get started with writing and running your tests without the need for complex setup or configuration files. Jest, on the other hand, often requires some configuration to get started, especially for more complex projects.

- **Bun APIs**: The test runner is built to work seamlessly with other Bun APIs. This native integration ensures that you can leverage the full power of Bun.

- **Isolated environments**: Each test file runs in an isolated environment in the Bun test runner. This isolation ensures that tests don't inadvertently affect each other, leading to more reliable and predictable test outcomes.

- **Snapshot testing**: Bun's test runner supports snapshot testing. This feature allows you to capture the *expected* state of your application and ensure that it doesn't change unexpectedly over time.

- **Watch mode**: Bun's test runner includes a watch mode that automatically reruns your tests when it detects changes in your code.

- **Native TypeScript support**: Given Bun's excellent support for TypeScript, its test runner naturally handles TypeScript out of the box. This is a significant advantage over Jest, where setting up TypeScript can be cumbersome and requires additional tools such as `ts-jest`.

- **Built-in coverage**: Bun's test runner comes with built-in support for test coverage, allowing you to track how much of your code is covered by tests.

Let's see how the test runner works in action.

Writing tests for our authentication endpoints

First, we will begin with tests for our authentication endpoints to validate that they execute correctly.

Registration tests

In this test, we are first going to instantiate our server and then check if our login and registration work with happy paths:

tests/auth.test.ts

```
import { beforeEach, describe, expect, test } from "bun:test";
import { createInMemoryApp } from "../src/controllers/main";
```

Here, we import a function to instantiate our in-memory application and utility functions from the testing module.

`describe` is going to be used as a wrapper for all our test cases where we can provide an additional description. Inside this wrapper, we are allowed to use additional hooks such as `beforeEach` and functions such as `test` and `expect`. Let's define the authentication tests:

```
describe("auth tests", () => {
  let app = createInMemoryApp();  beforeEach(async () => {
    app = createInMemoryApp();
  });
```

Here, we define our test suite, and we also instantiate our app. `beforeEach` is a helpful hook that gets executed before every test that helps us to recreate the instance of our app after every test run. It ensures that the data tests created are cleared after being executed. There are also similar hooks, such as `afterEach`, which executes after every test, and `beforeAll` and `afterAll`, which execute before and after the whole test suite, respectively.

Let's now define our first test case:

```
test("POST /register - normal case", async () => {
  const jsonBody = {
    email: "test@example.com",
    password: "password123",
    name: "Test User",
  };

  const response = await app.request("/api/v1/auth/register/", {
    method: "POST",
    headers: {
      "Content-Type": "application/json",
    },
    body: JSON.stringify(jsonBody),
  });

  expect(response.status).toBe(200);
});
```

`test` is a function that accepts a function to be executed during our test. In our test, we create the data required for registering the user, and then we use the `request` method on our application to make an endpoint call, which mimics a call we would get from the outside of our application.

We provide the `method` to the function and also the `Content-Type` header, which tells the server what our data is going to be.

Finally, we use the `expect` function, which makes sure that the `status` of the `response` is `200`. If the `status` is anything else, our test case will fail.

Let's now check that our endpoint works as expected when we try to register when the user exists and when we try to log in with a non-existent user.

We will start by registering a user from inside our test case.

tests/auth.test.ts

```
. . .
  test("POST /register - user already exists", async () => {    await
app.request("/api/v1/auth/register/", {
      method: "POST",
      headers: { "Content-Type": "application/json" },
      body: JSON.stringify({
        email: "existing@example.com",
        password: "password123",
        name: "Existing User",
      }),
    });
```

And then, we will try to register the user again:

```
    const response = await app.request("/api/v1/auth/register/", {
      method: "POST",
      headers: { "Content-Type": "application/json" },
      body: JSON.stringify({
        email: "existing@example.com",
        password: "password123",
        name: "Existing User",
      }),
    });
    expect(response.status).toBe(400);
  });
```

Here, we can expect that the status to be returned is 400, because the data we provided is no longer correct.

Let's move to the tests for our login endpoint.

Login tests

We will proceed with writing our tests with a case for our login endpoint. Let's start with the test for the successful case for login:

tests/auth.test.ts

```
  . . .
  test("POST /login - success", async () => {
    const res1 = await app.request("/api/v1/chat/", { method: "GET"
```

```
    });
        expect(res1.status).toBe(401);
```

Here, we check that when we try to access an endpoint that requires authentication without a correct token, we get 401.

We create a user that we will use for our login later on:

```
        await app.request("/api/v1/auth/register/", {
          method: "POST",
          headers: { "Content-Type": "application/json" },
          body: JSON.stringify({
            email: "loginuser@example.com",
            password: "password123",
            name: "Login User",
          }),
        });
```

We call the login endpoint with the data we used during the registration. We expect that now the login is successful, and we get 200:

```
        const loginResponse = await app.request("/api/v1/auth/login/", {
          method: "POST",
          headers: { "Content-Type": "application/json" },
          body: JSON.stringify({
            email: "loginuser@example.com",
            password: "password123",
          }),
        });
        expect(loginResponse.status).toBe(200);
        const token = (await loginResponse.json())["token"];
        expect(token).toBeTruthy();
```

Also, we retrieve the data of the response and deserialize it to JSON with a json() call to get the token out of it. Finally, we check that the token evaluates to a truthy value.

Now, we can check that the chat endpoint has become accessible:

```
        const res2 = await app.request("/api/v1/chat/", {
          method: "GET",
          headers: { Authorization: `Bearer ${token}` },
        });
        expect(res2.status).toBe(200);
    });
```

We provide an `Authorization` token and expect that the endpoint will let us in.

Let's proceed with the test case for login when the user doesn't exist:

```
...
  test("POST /login - non-existing user", async () => {
    const response = await app.request("/api/v1/auth/login/", {
      method: "POST",
      headers: { "Content-Type": "application/json" },
      body: JSON.stringify({
        email: "nonexisting@example.com",
        password: "password123",
      }),
    });
    expect(response.status).toBe(401);
  });
});
```

Here, we simply send credentials of non-existing users and rightfully expect that the login will not let us in.

Let's also add two test cases to check that Zod validation works correctly.

Validation tests

We will add tests to the same `describe` block that are going to check that our validation works as expected. We will try to send bodies with incorrect structures to both endpoints to see that the request doesn't get through and that we get the correct error messages. We will start by trying to register with an incorrect email structure and missing `password` and `name` fields:

```
test("POST /register - incorrect body", async () => {
  const jsonBody = {
    email: "example",
  };

  const response = await app.request("/api/v1/auth/register/", {
    method: "POST",
    headers: {
      "Content-Type": "application/json",
    },
    body: JSON.stringify(jsonBody),
  });

  expect(response.status).toBe(400);
```

We expect that the registration endpoint returns us 400 as we sent incorrect data, and this is what the status code points to. We can also observe that Zod provides us with detailed error data with specific errors:

```
expect(await response.json()).toEqual({
  success: false,
  error: {
    issues: [
      {
        validation: "email",
        code: "invalid_string",
        message: "Invalid email",
        path: ["email"],
      },
      {
        code: "invalid_type",
        expected: "string",
        received: "undefined",
        path: ["password"],
        message: "Required",
      },
      {
        code: "invalid_type",
        expected: "string",
        received: "undefined",
        path: ["name"],
        message: "Required",
      },
    ],
    name: "ZodError",
  },
});
});
```

We will use the `toEqual` method for deep equality checks, rather than the `toBe` method, which checks for strict equality (using ===). The `toEqual` method is more suitable for comparing objects, as it evaluates the equality of their contents.

The next test case we are going to cover is the validation for our login endpoint. We will similarly send incorrect data to the endpoint, and we will expect it to fail:

```
test("POST /login - incorrect body", async () => {
  const response = await app.request("/api/v1/auth/login/", {
    method: "POST",
```

```
    headers: { "Content-Type": "application/json" },
    body: JSON.stringify({
      email: "wrong",
    }),
  });
  expect(response.status).toBe(400);
```

The `email` field is in the wrong format and `password` is missing, so we will get a status of `400`. We can look at the response to check that the specific errors are correct:

```
  expect(await response.json()).toEqual({
    success: false,
    error: {
      issues: [
        {
          validation: "email",
          code: "invalid_string",
          message: "Invalid email",
          path: ["email"],
        },
        {
          code: "invalid_type",
          expected: "string",
          received: "undefined",
          path: ["password"],
          message: "Required",
        },
      ],
      name: "ZodError",
    },
  });
});
```

Now, you can run the tests with the following command:

```
$ bun test
```

You will be able to see that all the tests for authentication have run successfully.

We can now also write the tests that cover our chat endpoint.

Writing test for our chat endpoints

We also need to cover our chat and message endpoints, so this is what we are going to do now. You will mostly recognize all the code written in the tests, so I will only explain the bits that are new. We will start our file with the imports and the test setup:

tests/chat.test.ts

```
import { beforeEach, describe, expect, test } from "bun:test";
import { createInMemoryApp } from "../src/controllers/main";

describe("chat tests", () => {
  let app = createInMemoryApp();

  beforeEach(async () => {
    app = createInMemoryApp();
  });
```

Then, we are going to create a utility function, which we are going to use to create an authorization token:

```
async function getToken(email = "test@test.com"): Promise<string> {
  await app.request("/api/v1/auth/register/", {
    method: "POST",
    headers: { "Content-Type": "application/json" },
    body: JSON.stringify({
      email: email,
      password: "password123",
      name: "Chat User",
    }),
  });    const loginResponse = await app.request("/api/v1/auth/
login/", {
    method: "POST",
    headers: { "Content-Type": "application/json" },
    body: JSON.stringify({
      email: email,
      password: "password123",
    }),
  });
  const token = (await loginResponse.json()).token;
  return token!;
}
```

This test utilizes our register endpoint to first create a user and then our login endpoint to retrieve the token.

Let's proceed with the next utility function to create a chat:

```
async function createChat(token: string) {
  const createChatResponse = await app.request("/api/v1/chat/", {
    method: "POST",
    headers: {
      "Content-Type": "application/json",
      Authorization: `Bearer ${token}`,
    },
    body: JSON.stringify({ name: "Test Chat" }),
  });
  const response = await createChatResponse.json();
  const chatId = response.data.id;
  return chatId;
}
...
```

createChat is used to create a chat with the authorization token so that we can test whether the messages are correct. It uses the chat creation endpoint to achieve it.

Let's proceed with the test cases for our chats.

Chat tests

We will first check an easy scenario where a user can retrieve chats from the system. First, we will create a user and a chat using our utility functions, and then we are going to retrieve all the chats for the user:

tests/chat.test.ts

```
...
test("GET /chat/ - get user chats", async () => {
  const token = await getToken();
  const chatId = await createChat(token);
  const response = await app.request("/api/v1/chat/", {
    method: "GET",
    headers: { Authorization: `Bearer ${token}` },
  });
  expect(response.status).toBe(200);
  const responseData = await response.json();
  const data = responseData.data;
  expect(Array.isArray(data)).toBeTruthy();
  expect(data.length).toBe(1);
```

```
    expect(data[0].id).toBe(chatId);
  });
```

Here are the important parts of the preceding code block:

- First, we check that the response we get is 200.

- `expect(Array.isArray(data)).toBeTruthy();` confirms that the retrieved data is in an array format, as expected for multiple chats.

- `expect(data.length).toBe(1);` ensures that only one chat is returned for the user, matching the test setup where a single chat was created.

- `expect(data[0].id).toBe(chatId);` verifies that the ID of the returned chat matches the `chatId` of the chat created in the setup. This confirms that the correct chat is being retrieved for the user.

Let's now cover retrieving chat in a more complicated case. We will create multiple chats, and only some of them will belong to our user. Then we are going to check that the user gets only relevant chats.

First, we create two different users and two different chats:

```
test("GET /chat/ - get user chats when multiple chat and users are
available", async () => {
    const token = await getToken();
    const token2 = await getToken("email@email.com");
    const chatId = await createChat(token);
    const chatId2 = await createChat(token2);
```

Here, we can also test that we get different chats back depending on the token.

Now, we validate that we indeed get the relevant chats for every user:

```
    const response = await app.request("/api/v1/chat/", {
      method: "GET",
      headers: { Authorization: `Bearer ${token}` },
    });
    expect(response.status).toBe(200);
    const responseData = await response.json();
    const data = responseData.data;
    expect(Array.isArray(data)).toBeTruthy();
    expect(data.length).toBe(1);
    expect(data[0].id).toBe(chatId);

    const response2 = await app.request("/api/v1/chat/", {
      method: "GET",
      headers: { Authorization: `Bearer ${token2}` },
```

```
  });
  expect(response.status).toBe(200);
  const responseData2 = await response2.json();
  const data2 = responseData2.data;
  expect(Array.isArray(data2)).toBeTruthy();
  expect(data2.length).toBe(1);
  expect(data2[0].id).toBe(chatId2);
});
```

Let's proceed with the test cases for our messages.

Messages tests

In this test, we will create and get some chat messages to validate that message creation and retrieval work as expected.

Here, we have created a user, a chat, and a message that belongs to the chat:

tests/chat.test.ts

```
...
  test("POST, GET /chat/:id/message/ - create and get chat messages",
async () => {
    const token = await getToken();
    const chatId = await createChat(token);
    await app.request(`/api/v1/chat/${chatId}/message/`, {
      method: "POST",
      headers: {
        "Content-Type": "application/json",
        Authorization: `Bearer ${token}`,
      },
      body: JSON.stringify({ message: "Hello World" }),
    });
```

Now we can retrieve the messages for the chat:

```
    const response = await app.request(`/api/v1/chat/${chatId}/
message/`, {
      method: "GET",
      headers: { Authorization: `Bearer ${token}` },
    });

    expect(response.status).toBe(200);
    const messages = await response.json();
    expect(messages.data).toBeInstanceOf(Array);
```

```
    expect(messages.data.length).toBe(2);
    expect(messages.data[0].message).toBe("Hello World");
    expect(messages.data[1].message).toBe("dummy response");
});
```

The test validates that we indeed get the message belonging to the chat from our chat retrieval endpoint. As well as that, we see that we get an expected dummy message that mimics the response of our AI assistant.

Now, we can test that the validation works for our chat and messages endpoints too.

Chats validation tests

We will send incorrect data to chat and message creation endpoints to see that the endpoint doesn't proceed with the creation in this case. Let's start by sending the wrong data to the chat endpoint:

```
test("POST /chat - incorrect body", async () => {
  const token = await getToken();
  const jsonBody = {
    name: "",
  };

  const response = await app.request("/api/v1/chat/", {
    method: "POST",
    headers: {
      "Content-Type": "application/json",
      Authorization: `Bearer ${token}`,
    },
    body: JSON.stringify(jsonBody),
  });
```

Here, we create a chat with a name that's too short, so we expect that the test is going to fail. Now, we can validate that it does indeed fail and see what we get in response:

```
expect(response.status).toBe(400);
expect(await response.json()).toEqual({
  success: false,
  error: {
    issues: [
      {
        code: "too_small",
        minimum: 1,
        type: "string",
        inclusive: true,
```

```
                        exact: false,
                        message: "String must contain at least 1 character(s)",
                        path: ["name"],
                    },
                ],
                name: "ZodError",
            },
        });
    });
```

We check that the response errors are the ones we expect to see from Zod here.

Let's proceed with the last test case for message validation:

```
test("POST /chat/:id/message - incorrect body", async () => {
    const token = await getToken();
    const response = await app.request(`/api/v1/chat/a/message/`, {
        method: "POST",
        headers: {
            "Content-Type": "application/json",
            Authorization: `Bearer ${token}`,
        },
        body: JSON.stringify({}),
    });
```

Here, we provide an incorrect chat ID in the URL and we send an empty body to the endpoint, so we expect it to fail now.

Our test cases fail with 400, and Zod provides the intended response that our message field is required:

```
    expect(response.status).toBe(400);
    expect(await response.json()).toEqual({
        success: false,
        error: {
            issues: [
                {
                    code: "invalid_type",
                    expected: "string",
                    received: "undefined",
                    path: ["message"],
                    message: "Required",
                },
            ],
            name: "ZodError",
```

```
      },
    });
  });
});
```

With all of this, we can run all our tests and see how much we have covered with the following command:

```
$ bun test --coverage
```

You will see a table that shows that we have covered more than 90% of our code lines, which means that during our tests, 90% of our code lines were executed. This is a great accomplishment, and now we can be much surer that our code has no bugs.

Summary

In this chapter, we've significantly reinforced the stability and reliability of our chat application's backend. By integrating Zod, we've introduced stringent validation rules, ensuring that data flowing into our endpoints meets our precise specifications.

Then, we used Bun's testing module, delving into writing comprehensive tests that scrutinize facets of our application's functionality. Together, Zod's validation and Bun's testing framework form a formidable duo, safeguarding our application against unexpected behaviors and vulnerabilities.

As we progress, these tools will be invaluable in maintaining the quality and reliability of our backend, ensuring that our application not only meets but exceeds the demands of real-world development. The next steps will involve focusing on more advanced aspects of backend development such as logging, security, caching, and debugging.

6

Advanced Backend Development – Security, Throttling, Caching, and Logging

In the previous chapter, we focused on validation and testing, which significantly improved the reliability of our server. With this, we have almost finished the development of our REST API, and we can now focus on the more advanced aspects such as **security**, **request throttling**, **caching**, and **logging** techniques. First, we will address potential security issues that our backend doesn't protect us from yet, and we will also add a layer of protection against **DoS attacks** with request throttling. Then, we will focus on how to cache the response we produce and configure and use logging in our application. With this in place, we will make sure that our application is secure, quick, and easy to debug.

In this chapter, we are going to cover the following topics:

- Managing security aspects
- Adding request throttling
- Managing cache
- Using logging

We will begin with the security aspects, focusing on what kind of attacks are possible in the context of a REST API and what we need to do to protect our server against them.

Technical requirements

For this chapter, we will need to add the Pino library for our logging. We can add it to our project by running the following command in the terminal:

```
$ bun add pino
```

All the code examples we will discuss are available in the GitHub repository: `https://github.com/PacktPublishing/Full-Stack-Web-Development-with-TypeScript-5/tree/main/Chapter06`.

Managing security aspects

So far, our app has a few security holes that we need to address before it can face real users. Making sure that it's not hackable and reducing the risk of potential vulnerabilities is essential for any web service, so let's focus first on which type of common and dangerous attacks can be executed against our REST server and how we can protect against them:

- **SQL injection**: SQL injection is a type of attack where the attacker manipulates SQL queries by injecting malicious SQL code through the REST endpoint input data. This can happen when user input is directly used in constructing SQL queries without proper validation or escaping. An attacker might exploit this vulnerability to access, modify, or delete data in a database, potentially gaining unauthorized access to sensitive information or even taking control of the database. To prevent SQL injection, we need to always use prepared statements and parameterized queries, validate and sanitize all user inputs, and apply least privilege access controls to the database.

- **XSS attack**: An XSS attack occurs when an attacker injects malicious scripts into content that is served to other users. These scripts execute within the context of the victim's browser under the trust level of the web application, allowing the attacker to steal cookies, session tokens, or other sensitive information reflected in the web browser. XSS can be performed by including malicious JavaScript in user-generated content that is not properly sanitized by the server or the client before being presented to other users. Defending against XSS involves encoding and escaping user input, implementing a **Content Security Policy** (**CSP**), and validating all input data. A CSP is typically handled on the frontend by escaping the user-generated content before putting it into the HTML, but it doesn't hurt to additionally address it on the backend either.

- **DoS attack**: A DoS attack aims to make our server unavailable to its intended users by overwhelming it with a flood of requests. This can be achieved through various means, such as sending more requests than the server can handle or exploiting a vulnerability that causes the server to crash. An attacker might send rapid, large, or complex requests to the API endpoints to exhaust server resources. Protection measures include rate limiting, filtering traffic to identify and block malicious patterns, and deploying DoS protection tools or services such as Cloudflare.

- **Unauthorized domain request**: This security concern involves making requests to a REST API from a domain that is not authorized by the API server's **Cross-Origin Resource Sharing (CORS)** policy. CORS is a mechanism that allows or restricts resources on a web server to be requested from another domain. Without proper CORS settings, an attacker could make unauthorized API calls from a malicious domain, potentially exposing sensitive information or exploiting vulnerabilities. To mitigate this risk, we need to configure CORS policies to explicitly allow only trusted domains to make requests and use other security measures, such as API keys and OAuth tokens, for authentication and authorization.

- **Man-in-the-middle attacks**: In a man-in-the-middle attack, an attacker intercepts the communication between a client and a server to eavesdrop on or alter the data being exchanged. This could compromise the confidentiality and integrity of the data, allowing the attacker to steal sensitive information or inject malicious content. HTTPS with TLS encryption is essential for protecting against man-in-the-middle attacks by ensuring that data in transit is encrypted and authenticated.

To summarize, to protect against these attacks, we will need to add HTTPS support, sanitize user input, configure CORS, and add rate limiting. How the first one is handled depends on how we are going to deploy our app, typically using the proxy web server or the load balancer that we are going to put in front of our backend application, and the second one will be addressed when we integrate with database data sources in the next chapter. Rate limiting is typically also implemented by the load balancer, but we are going to implement it in this chapter for illustration purposes. We are also going to configure CORS settings for our backend server. Let's start with configuring CORS.

The CORS setup basically tells our server which request methods to accept, which domain to accept them from, and for how long we should cache the CORS query for the caller. Let's configure CORS by using the built-in hono library – `hono/cors`. I will only list here the code that we need to add to include CORS, indicating omitted code that is going to remain the same with . . . .

First, we are going to provide configuration for the CORS setup, and then we are going to add it as middleware for our app:

src/controllers/main.ts

```
import { cors } from "hono/cors";
...
const corsOptions = {
  origin: [ Bun.env.CORS_ORIGIN as string],
  allowMethods: ["GET", "POST", "PUT", "PATCH", "DELETE"],
  allowHeaders: ["Content-Type", "Authorization"],
  maxAge: 86400,
};
...
export function createMainApp(
```

```
   authApp: Hono<ContextVariables>,
   chatApp: Hono<ContextVariables>,
) {
...
   app.use("*", cors(corsOptions));s


...

}
```

Let's discuss what is happening here. First, we import the `cors` library. Then, we configure `corsOptions`. In `corsOptions`, these are the following configurations:

- `origin`: Restricts which domains are allowed to make requests to the server. For it to work, add `CORS_ORIGIN=http://localhost:5173` to your `.env` file. Therefore, only requests originating from `http://localhost:5173` are permitted.

- `allowMethods`: Lists the HTTP methods that are allowed when accessing the resource. In this case, the server accepts the `GET`, `POST`, `PUT`, `PATCH`, and `DELETE` methods from the allowed origin.

- `allowHeaders`: Specifies the headers that can be included in the requests made to the server. `Content-Type` and `Authorization` are explicitly allowed, facilitating the use of content types such as JSON and authorization mechanisms such as tokens or basic authentication.

- `maxAge`: Indicates how long (in seconds) the results of a preflight request can be cached. Here, it is set to `86400` seconds, or `24` hours, meaning the browser can cache the preflight response for a day before needing to send another preflight request for subsequent requests.

  Lastly, we add the `cors` function with our configuration as middleware to our main `hono` app. We can only send requests to our server from `localhost:5173`, which is supposed to be our frontend local development URL and which we will need to change to the real domain URL in production. We also allow the common HTTP methods and the headers that we are using to proceed to our endpoint handler.

Now, we will discuss request throttling.

Adding request throttling

Request throttling or **rate limiting** is a technique that helps us against DoS attacks. It essentially counts the number of requests per user and doesn't allow the user to perform too many requests in a given time frame. Typically, we would also identify the IP of the caller to protect our non-authorized endpoints, but as we cannot get such a low-level detail of the connection in `hono`, we will focus on how to implement middleware that will do request throttling of the authorized endpoints based on `userId`.

Writing the middleware

Let's write middleware in which we will count how many requests a user has made in the last 15 minutes and throw a 429 status code, which means a user has made more than 100 requests:

src/middlewares/rateLimiting.ts

```
import type { Context } from "hono";
import type { ContextVariables } from "../constants";

const requestCounts = new Map<string, { count: number; resetTime:
number }>();

const MAX_REQUESTS = 100; // Max requests per window per client
const WINDOW_SIZE_MS = 15 * 60 * 1000; // 15 minutes in milliseconds
export const rateLimitMiddleware = async (
  c: Context<ContextVariables>,
  next: Function,
) => {
  const userId = c.get("userId");
  if (!userId) {
    await next();
    return;
  }
  const now = Date.now();
  let requestData = requestCounts.get(userId);
...
```

Let's discuss what is happening here, line by line:

- `const requestCounts = new Map<string, { count: number; resetTime: number }>();`: Initializes a map to keep track of request counts and reset times for each user, indexed by `userId`

- `const MAX_REQUESTS = 100;`: Defines the maximum number of requests a user is allowed to make within the specified window, `WINDOW_SIZE_MS`

- `const WINDOW_SIZE_MS = 15 * 60 * 1000;`: Sets the duration of the rate-limiting window to 15 minutes, converted to milliseconds

- `export const rateLimitMiddleware = async (c: Context<ContextVariables>, next: Function) => { ... };`: Defines the asynchronous middleware function, `rateLimitMiddleware`, which takes Hono's `context` object and a `next` function to proceed to the next middleware

- `const userId = c.get("userId");`: Attempts to retrieve the `userId` from the context, which is expected to be set by previous middleware in the request handling pipeline

- `if (!userId) { await next(); return; }`: If no `userId` is found, the middleware immediately proceeds to the next middleware without applying rate limiting

- `let requestData = requestCounts.get(userId);`: Retrieves the current request data for the user from the `requestCounts` map

Implementing the middleware

Let's proceed with the implementation:

src/middlewares/rateLimiting.ts

```
...
  if (!requestData) {
    requestData = { count: 1, resetTime: now + WINDOW_SIZE_MS };
    requestCounts.set(userId, requestData);
  } else {
    if (requestData.resetTime < now) {
      requestData.count = 1;
      requestData.resetTime = now + WINDOW_SIZE_MS;
    } else {
      requestData.count += 1;
    }
  }

  if (requestData.count > MAX_REQUESTS) {
    return c.text("Rate limit exceeded. Try again later.", 429);
  } else {
    requestCounts.set(userId, requestData);
    await next();
  }
};
```

Let's discuss what we see here:

- `if (!requestData) { requestData = { count: 1, resetTime: now + WINDOW_SIZE_MS }; requestCounts.set(userId, requestData); }`: If there's no existing data for the user, this initializes it with a count of 1 and sets the reset time to 15 minutes from the current time.

- `if (requestData.resetTime < now) { requestData.count = 1; requestData.resetTime = now + WINDOW_SIZE_MS; } else { requestData.count += 1; }`: Checks whether the current time is past the reset time. If so, it resets the count and the reset time. If not, it increments the request count.

- `if (requestData.count > MAX_REQUESTS) { return c.text("Rate limit exceeded. Try again later.", 429); }`: If the user has exceeded the maximum number of allowed requests, this returns a 429 status code (*Too Many Requests*) and a message indicating that the rate limit has been exceeded.

- `else { requestCounts.set(userId, requestData); await next(); }`: If the rate limit has not been exceeded, this updates the user's data in the `requestCounts` map and calls `next()` to proceed to the next middleware.

Including the middleware in the main app

Now, we can include `rateLimitMiddleware` in the main app as middleware in a similar way we included our `cors`. I will only show the changes in the file; everything else is the same:

src/controllers/main.ts

```
import { rateLimitMiddleware } from "../middlewares/rateLimiting";
...
export function createMainApp(
  authApp: Hono<ContextVariables>,
  chatApp: Hono<ContextVariables>,
) {
...
  app.use("*", attachUserId);
  app.use("*", rateLimitMiddleware);
...

}
```

Here, we simply add the newly created middleware to our main app.

Now, our backend has a proper configuration to protect us against requests from unauthorized domains and DoS attacks, and we are ready to discuss how we can manage cache in our app.

Managing the cache

Cache is essential to speed up our endpoints, as most of the user requests are going to return the same data on the GET endpoints. Cache is typically used to avoid load-intensive code pieces from executing again.

In this part, we are going to implement caching middleware that can be used from inside our endpoints to add and remove elements from the cache. One trick with caching is **cache invalidation**, which is a topic on its own, as when we create a new chat, for example, the cache value to get all chats is no longer valid, so we need to remove it. So, we are going to add middleware that provides this functionality, as well as associate cache values with userId.

Writing the cache middleware

Here, we will define middleware that will cache the results of the calculations of the endpoint calls, which we can reuse again:

src/middlewares/cacheMiddleware.ts

```
import type { Context } from "hono";
import type { ContextVariables } from "../constants";

interface CacheEntry {
  body: any;
  expiration: number;
}
```

The interface defines the structure for cache entries, including the cached body and expiration timestamp.

The following code lines initialize an in-memory cache to store the responses, keyed by a combination of the request path and userId:

```
export const cacheMiddleware = () => {
  const cache = new Map<string, CacheEntry>();
```

In the next lines, we will define our middleware function:

```
  return async (c: Context<ContextVariables>, next: () =>
Promise<void>) => {
    const userId = c.get("userId");
    const path = c.req.path;
    const cacheKey = `${path}:${userId}`;
```

In the preceding code block, we have the following lines:

- `const userId = c.get("userId");`: Retrieves `userId` from the context
- `const path = c.req.path;`: Extracts the request path from the context's request object, which will identify which endpoint we are using
- `const cacheKey = ${path}:${userId};`: Constructs a unique cache key using the request path and `userId`, ensuring cache entries are unique per user and request path

The following code snippet attaches a cache object to the context with methods to cache responses, clear the cache for the current path and user, and clear the cache for any specified path and the current user:

```
c.set("cache", {
  cache: (body: object, expiration: number = 3600) => {
    const expireAt = Date.now() + expiration * 1000;
    const entry = { body, expiration: expireAt };
    cache.set(cacheKey, entry);
  },
```

The following method is used to delete the cached data for the default key:

```
  clear: () => {
    cache.delete(cacheKey);
  },
```

Here, we delete cached data for a passed key:

```
  clearPath: (path: string) => {
    const fullKey = `${path}:${userId}`;
    cache.delete(fullKey);
  },

});
```

In the next lines, we retrieve the cache when needed:

```
if (c.req.method.toUpperCase() === "GET") {
  const cacheEntry = cache.get(cacheKey);
  if (cacheEntry) {
    if (cacheEntry.expiration > Date.now()) {
      return c.json(cacheEntry.body);
    }
  }
}
```

```
    await next();
  };
};
```

Let's discuss these code lines:

- `if (c.req.method.toUpperCase() === "GET") { ... }`: Checks whether the request method is GET before attempting to retrieve data from the cache, implying that only GET requests are considered for caching.

- `const cacheEntry = cache.get(cacheKey);`: Attempts to retrieve a cache entry using the constructed cache key.

- `if (cacheEntry && cacheEntry.expiration > Date.now()) { return c.json(cacheEntry.body); }`: If a valid cache entry is found (i.e., it hasn't expired), this returns the cached response immediately without proceeding to subsequent middleware or handlers.

- `await next();`: If no valid cache entry is found, or the request is not a GET request, control is passed to the next middleware or handler in the chain, potentially to fetch fresh data and handle the request.

As we are adding a new variable to our context, we also need to change our `ContextVariables` to adapt to the change by providing a new `cache` key, which is going to define the interface of the function we will use from our endpoints:

src/constants.ts

```
export const API_PREFIX = "/api/v1";

export type ContextVariables = {
  Variables: {
    userId: string;
    cache: {
      cache: (body: object, expiration?: number) => void;
      clear: () => void;
      clearPath: (path: string) => void;
    };
  };
};
```

Here, we provide a new `cache` key that exposes the methods we defined in the *Writing the cache middleware* section.

Using the cache middleware in our chat endpoints

Now, let's see how we can use cache implementation in our chat endpoints to cache the results of getting all the lists, allowing us to simply return the cache when we are asked to again. We will also invalidate this cache when we create a new chat. We will also implement similar logic for chat messages. You've seen most of the code already, so I'm going to explain only the code bits relevant to caching. We will utilize our cache object attached by the middleware in our endpoints to set and retrieve the cache data we return. We will cache the responses of the GET endpoints and clear the cache of the POST endpoints. I am going to highlight the usage of the cache we introduced:

src/controllers/chat.ts

```
...
export function createChatApp(
  chatResource: IDatabaseResource<DBChat, DBCreateChat>,
  messageResource: IDatabaseResource<DBMessage, DBCreateMessage>,
) {
  const chatApp = new Hono<ContextVariables>();

  chatApp.post(CHAT_ROUTE, zValidator("json", chatSchema), async (c)
=> {
    const userId = c.get("userId");
    const { name } = c.req.valid;
    const data = await chatResource.create({ name, ownerId: userId });
    c.get("cache").clearPath(c.req.path);
    return c.json({ data });
  });
```

We clear the cache to get all the chats when we create a new one.

Here, we cache the result of retrieving all the chats:

```
  chatApp.get(CHAT_ROUTE, async (c) => {
    const userId = c.get("userId");
    const data = await chatResource.findAll({ ownerId: userId });
    const res = { data };
    c.get("cache").cache(res);
    return c.json({ data });
  });
```

Next, we cache the result of the individual chat retrieval:

```
chatApp.get(CHAT_DETAIL_ROUTE, zValidator("param", idSchema), async
(c) => {
  const { id } = c.req.valid("param");
  const userId = c.get("userId");
  const data = await chatResource.find({ id, ownerId: userId });
  const res = { data };
  c.get("cache").cache(res);
  return c.json({ data });
});
```

Then, we set the cached data for all the messages after retrieval:

```
chatApp.get(CHAT_MESSAGE_ROUTE, zValidator("param", idSchema), async
(c) => {
  const { id: chatId } = c.req.valid("param");
  const data = await messageResource.findAll({ chatId });
  const res = { data };
  c.get("cache").cache(res);
  return c.json(res);
});
```

And finally, we clear the cache for our chat messages when we create two new ones:

```
chatApp.post(
  CHAT_MESSAGE_ROUTE,
  zValidator("param", idSchema),
  zValidator("json", messageSchema),
  async (c) => {
    const { id: chatId } = c.req.valid("param");
    const { message } = c.req.valid("json");

    const userMessage: DBCreateMessage = { message, chatId, type:
"assistant" };
    await messageResource.create(userMessage);

    const responseMessage: DBCreateMessage = {
      message: "dummy response",
      chatId,
      type: "user",
    };
```

```
      const data = await messageResource.create(responseMessage);
      const res = { data };
      c.get("cache").clearPath(c.req.path);
      return c.json(res);
    },
  );
  return chatApp;
}
```

Now, our requests are GET requests that are cached for 15 minutes, and we don't need to put the load on our data sources to retrieve the data again if it doesn't change. We also ensure that when the data is obsolete, the cache is invalidated.

Let's now turn to logging, how to configure it, and how we can use it on the example of our two newly created middlewares.

Using logging

Logging is an essential technique that helps us to reconstruct what happened on our backend and also know the other important events that happened. It's very useful when we actually debug our code to understand what went wrong, and it's helpful in trying to get more context. In our case, we will use the **pino** logger, which provides useful functionality out of the box and also allows a decent level of configuration.

Let's first create a general configuration for our logger so that we can use it from other parts of our application.

Creating our logger's configuration

We will initialize a main logger with a logging level and then export it:

src/loggers.ts

```
import pino from "pino";

const mainLogger = pino({
  level: Bun.env.LOG_LEVEL || "info",
  timestamp: pino.stdTimeFunctions.isoTime,
});

export default mainLogger;
```

Here, we instantiate a main logger, where we specify which level of logging to track and also whether to include a timestamp to the logs in the ISO format. If you want a different level of logging, you can expand the .env file with a LOG_LEVEL variable and set the variable to debug if you want to see more fine-grained logging.

Now, we can utilize this logger in other parts of our system, and when we execute the app, we will see it output to standard output.

Adding logger to our caching middleware

Let's add logging to our caching middleware so that we can observe what is happening there. I'm going to explain the logging-relevant lines, as other things are just going to remain unchanged. First, we are going to create a child logger of our main logger, and then we are going to write a log line with it when we set and retrieve the cache:

src/middlewares/cacheMiddlewares.ts

First, we import the main logger we created in the *Creating our logger's configuration* section:

```
import mainLogger from "../logger";
```

Next, we create a child logger of the main logger. The child logger adds a new property name to every new log statement, which is going to identify the logs that relate to the cache:

```
const logger = mainLogger.child({ name: "cacheMiddleware" });
```

Next, we will see the untouched logic of our cache middleware until the first log line:

```
export const cacheMiddleware = () => {
  const cache = new Map<string, CacheEntry>();

  return async (c: Context<ContextVariables>, next: () =>
Promise<void>) => {
    const userId = c.get("userId");
    const path = c.req.path;
    const cacheKey = `${path}:${userId}`;

    c.set("cache", {
      cache: (body: object, expiration: number = 3600) => {
        const expireAt = Date.now() + expiration * 1000;
        const entry = { body, expiration: expireAt };
```

Until the next line in the following code block, everything is untouched, and now, we will add a log statement that sets a new cache key.

When we clear cache, we will log this action as well:

```
    logger.info(
      `Setting cache key: ${cacheKey}, to${JSON.
stringify(entry)}`,
    );
    cache.set(cacheKey, entry);
  },
  clear: () => {
    logger.info(`Clearing cache key: ${cacheKey}`);
```

When we clear the specific path, we will log that as well:

```
    cache.delete(cacheKey);
  },
  clearPath: (path: string) => {
    const fullKey = `${path}:${userId}`;
    logger.info(`Clearing cache key: ${fullKey}`);
```

We log information when we encounter a cache entry:

```
    cache.delete(fullKey);
  },
});

if (c.req.method.toUpperCase() === "GET") {
  const cacheEntry = cache.get(cacheKey);
  if (cacheEntry) {
    logger.debug(
      `Found cache entry: ${cacheKey}, to${JSON.
stringify(cacheEntry)}`,
    );
```

We also log that we actually use the cache so that we know we didn't execute the endpoint:

```
    if (cacheEntry.expiration > Date.now()) {
      logger.debug(
        `return from key: ${cacheKey}, body: ${JSON.stringify(
          cacheEntry.body,
        )}`,
      );
```

We also add a log statement when our cache entry expires:

```
      return c.json(cacheEntry.body);
    } else {
      logger.debug(
        `Cache entry expired cache key: ${cacheKey}, expiration:
${cacheEntry?.expiration}`,
      );
```

Logging is essential, as it is going to be one of the tools that will help us to fix issues and see what happens in our system. With this, we are ready to focus on our data storage in more detail in the next chapter.

Summary

In this chapter, we significantly improved the security, speed, and trackability of our service. By integrating a CORS setup and request throttling, we made sure that our app is secure and can face real users, while caching helps us execute our app promptly and use as few resources as possible. In addition, we learned how to make it easier to debug our application with the use of logging.

In the next chapter, we will move on to improve our data storage, and we will also deal with the issue of our data disappearing every time we reload our backend application, with the use of an actual **PostgreSQL** database.

Part 3: Integrating PostgreSQL for Data Management

In this part, you will learn how to effectively integrate and manage databases using PostgreSQL with TypeScript. It includes setting up PostgreSQL, using libraries to interact with the database, and utilizing **Object-Relational Mappings (ORMs)** for efficient data management. This part is crucial for understanding how to handle data in a robust and scalable manner, essential for any full-stack development project.

This part includes the following chapters:

- *Chapter 7, PostgreSQL Basics, Storage, and Setup*
- *Chapter 8, Interacting with PostgreSQL Using Libraries*
- *Chapter 9, Interacting with PostgreSQL Using Prisma ORM*

7

PostgreSQL Basics, Storage, and Setup

After mastering the advanced aspects of backend development, we're ready to shift our focus to a key element of any dynamic application: persistent storage. This transition moves us into database management, leveraging PostgreSQL and Docker for reliable data storage and access.

In this chapter, we'll dive into the implementation of persistent storage essential for web development, ensuring data persistence across server restarts. Our learning path includes discussing Docker's role in creating and managing containerized applications, detailing steps to deploy PostgreSQL within a Docker container, designing a schema for our chat application, and interacting with the database through **Create, Read, Update, and Delete (CRUD)** operations.

Understanding persistent storage is fundamental for real-world applications, ensuring data such as messages and user details remain accessible and intact. By the end of this chapter, you'll possess both theoretical knowledge and practical experience in creating a database for a chat application using PostgreSQL. We are going to cover the following topics:

- Setting up PostgreSQL in Docker

- Constructing the database schema

- Writing CRUD Structured Query Language (SQL) operations

First, we will install Docker, and then we will create a container for our PostgreSQL instance using Docker. Let's start with the installation.

Technical requirements

To proceed with this chapter, we will need to install Docker on our system. The simplest option to get it for Mac, Linux, and Windows is to install Docker Desktop, which is going to provide us with the required Docker Engine and Docker Compose.

> **Important note**
> Be careful with the Docker Desktop license as it's paid for commercial use.

You can install Docker Desktop by following the official guide from Docker based on your platform: `https://docs.docker.com/desktop/`.

Alternatively, if you are on Linux, you can install Docker Engine and Docker Compose easily without Docker Desktop by following these links:

`https://docs.docker.com/engine/install/`

`https://docs.docker.com/compose/install/`

All the code we are going to discuss in this chapter is available at `https://github.com/PacktPublishing/Full-Stack-Web-Development-with-TypeScript-5/tree/main/Chapter07`.

Setting up PostgreSQL in Docker

Now that we have Docker installed on our system, we can set up a PostgreSQL instance using Docker on our computer, but before we do it, let's discuss what Docker is and what it is useful for in the context of web development.

What are Docker and Docker Compose?

Docker is a platform that allows you to package your application and its dependencies into a container, which can be easily shipped and run in any environment and any operating system in a similar manner. Think of it like this: imagine your application is a delicate piece of furniture that needs to be transported from one place to another. Docker acts as the perfect shipping container for your furniture. It carefully wraps up your application and all its dependencies, making sure everything is securely packaged together. No matter where the shipping container ends up, whether it's a different room, a different house, or even a different country, you can be confident that when you open the container, your furniture will be intact and ready to use.

Docker ensures that your application can be seamlessly moved and deployed across different environments without worrying about compatibility issues or missing pieces. This makes the setup easier to develop in a team that runs different operating systems as well, and it simplifies the deployment as a lot of infrastructure platforms support running a Docker container.

Docker containers are streamlined, self-sufficient packages capable of running applications by including all necessary components such as code, runtime environment, system tools, libraries, and settings. This ensures that your application will run the same way, regardless of where it's deployed, solving the classic *it works on my machine* problem.

Docker Compose, meanwhile, is a utility for orchestrating applications that use multiple Docker containers. It allows you to use a YAML file to set up and link your application's services, networks, and storage, simplifying the process of managing complex container setups. Once the configuration is done, you can create and start all the services from your configuration with a single command. It simplifies the development process by allowing us to define a complex stack of services (such as web servers, databases, cache services, and so on) that make up an application in a straightforward and declarative manner.

In web development, Docker and Docker Compose are incredibly useful for several reasons:

- **Consistency**: Docker ensures that your application runs in the same environment during development, testing, and production. This consistency reduces bugs and improves quality.

- **Microservices architecture**: Docker is ideal for microservices architecture, where different parts of an app are developed and deployed independently. It allows each microservice to be containerized with its dependencies.

- **Isolation**: Containers are isolated from each other and the host system. This isolation enhances security and allows you to run multiple containers on the same host without dependencies conflicting.

- **Scalability**: With Docker, you can easily scale up or down by simply starting more or fewer containers without affecting the application.

- **Database management**: Docker can be used to run database services in containers. This approach simplifies database setup, backups, replication, and scaling. It also ensures that developers are working with the same database configuration, reducing environment discrepancies and allowing us to scrap and recreate the database easily.

Now that we understand better what Docker and Docker Compose are, let's talk about our choice of PostgreSQL as a database.

What is PostgreSQL?

PostgreSQL, often simply called **Postgres**, is an advanced, open source **relational database management system (RDBMS)** prized for its reliability, flexibility, and adherence to technical standards. It's built to manage various tasks, from running on individual computers to powering data warehouses or web services that support numerous simultaneous users.

Here are the main benefits of PostgreSQL:

- **Advanced features**: PostgreSQL includes a wide array of advanced features out of the box, such as complex queries, window functions, foreign data wrappers, and support for storing and querying JSON and XML data. These features enable us to handle complex data workloads and patterns directly within the database.

- **Extensibility**: One of PostgreSQL's standout features is its extensibility. We can define our own data types, triggers, and custom functions.

- **Standards compliance**: PostgreSQL has a strong emphasis on SQL standards compliance. This adherence ensures that applications built on PostgreSQL can be easily ported to other SQL-compliant databases, providing greater flexibility and future-proofing for businesses.

- **Performance and reliability**: PostgreSQL offers sophisticated optimization features for complex queries, robust transaction management, and fault tolerance through features such as point-in-time recovery, tablespaces, and replication. PostgreSQL can handle large volumes of data with high concurrency, making it suitable for enterprise-level applications.

- **Strong community and support**: PostgreSQL has a vibrant, active community, contributing to its continuous development and support. This community provides a wealth of resources, including extensive documentation, third-party tools, and active forums for troubleshooting and advice.

- **Cost-effectiveness**: Being open-source, PostgreSQL is free to use. This makes it an attractive option for start-ups and companies looking to reduce their operational costs without compromising on the quality and capabilities of their **database management system (DBMS)**.

Here is how PostgreSQL compares to other database solutions:

- **MySQL/MariaDB**: PostgreSQL is frequently compared to MySQL or its fork, MariaDB, which are also popular open source RDBMSs. While MySQL is renowned for its speed and reliability in read-heavy scenarios, PostgreSQL shines with its advanced features such as complex queries and support for **multiple concurrent transactions** (through **multi-version concurrency control**, or **MVCC**). PostgreSQL's extensibility and standards compliance, including full ACID compliance for transactions, make it a preferred choice for complex and mission-critical applications. MVCC and ACID are more advanced database concepts, but if you haven't heard of them, I recommend reading up on them when you feel more comfortable with databases.

- **SQLite**: SQLite is a lightweight, file-based database. It's designed for simplicity and minimal setup, making it ideal for embedded applications and small projects. PostgreSQL, in contrast, offers more robustness and scalability, supporting large datasets and concurrent users more effectively.

- **NoSQL databases**: NoSQL databases such as MongoDB or Cassandra offer schema flexibility and scalability, particularly for unstructured data. PostgreSQL, while primarily a relational database, also incorporates JSON support and other NoSQL features, allowing for both structured and unstructured data management in a single system. This hybrid approach, combined with its reliability and ACID compliance, makes PostgreSQL a versatile choice for a wide range of applications.

In general, PostgreSQL is slowly but steadily becoming a de facto standard for web development and new applications due to all the features, speed, and flexibility it provides.

We are now ready to use Docker to create our PostgreSQL server.

Creating a database as a Docker container

To create our PostgreSQL server in Docker, we will use Docker Compose with Compose files. We will define our PostgreSQL service by using an official PostgreSQL Docker image, which serves as the preset with all commands required to create a PostgreSQL in a new sandbox container. Then, we will make the database accessible from our local host.

First, we are going to define which base image we are going to use, then pass environment variables and set up how to access our database from the outside:

docker_compose_test.yml

```
version: '3.8'
services:
  postgres:
    image: postgres:latest
    restart: always
    environment:
      POSTGRES_DB: test
      POSTGRES_USER: test
      POSTGRES_PASSWORD: test
    ports:
      - "5434:5432"
    volumes:
      - data/postgres_test:/var/lib/postgresql/data
```

Let's discuss what every line does in the preceding `docker_compose_test.yml` file:

- `version`: Specifies the version of the Docker Compose file syntax. `3.8` is one of the latest versions supporting most Docker features.

- `services`: Defines the containers that make up your application. Here, it's defining a single service named `postgres`.

- `image`: Specifies the Docker image to use for the container. `postgres:latest` pulls the latest official PostgreSQL image from Docker Hub.

- `restart`: Configures the restart policy for the container. `always` means the container will restart automatically if it stops.

- `environment`: Sets environment variables in the container. This configuration specifies the default database (`POSTGRES_DB`), user (`POSTGRES_USER`), and password (`POSTGRES_PASSWORD`) for PostgreSQL, which we will need to connect to the database.

- ports: Maps ports from the container to the host machine. By default, the network of the Docker container has nothing to do with our local networking; it has its own dedicated network, so we need a way to get in, which is what ports are for. 5434:5432 maps the default PostgreSQL port inside the container (5432) to port 5434 on the host, allowing access to the database on localhost:5434 from our computer.

- volumes: Persists data generated by and used by Docker containers. Here, postgres_data_test:/var/lib/postgresql/data maps a named volume (postgres_data_test) to the data directory inside the container, ensuring that database data persists across container restarts by the given path.

We are going to use the database created by docker_compose_test.yml for testing so that we can safely create and delete data in this database without the risk of messing up with our main database.

Here is the Docker Compose setup for our main development database:

docker_compose.yml

```
version: '3.8'
services:
  postgres:
    image: postgres:latest
    restart: always
    environment:
      POSTGRES_DB: db
      POSTGRES_USER: user
      POSTGRES_PASSWORD: pass
    ports:
      - "5433:5432"
    volumes:
      - ./data/postgres:/var/lib/postgresql/data
```

The content of the docker_compose.yml file is almost the same as in our test docker-compose file, but it exposes a different port – 5433 – so it's accessible at http://localhost:5433, uses different credentials, and stores its volume data at a slightly different path in our filesystem.

Now, we can run our PostgreSQL service in a container through this command:

```
$ docker-compose -f docker_compose_test.yml up
```

> **Important note**
>
> If you get a . . . is not shared from the host and is not known to docker
> error in Docker Desktop on Mac, you need to add the path of your project to the File Sharing
> config. You can do it by adding the project path to Docker Desktop >> Preferences
> >> Resources >> File Sharing.

You should be able to see that the database image is pulled and then that it is being briefly set up, which will conclude with the following similar message in your terminal:

```
postgres_1  | 2024-02-18 16:47:36.724 UTC [1] LOG:  database system is
ready to accept connections
```

Now, our database is accessible at localhost:5434 and is ready to be interacted with.

Next, it is time to define the schema of our database.

Constructing the database schema

A database structure, or schema, is a blueprint that defines how data is organized in a database. It includes tables, columns within those tables, and the types of data that can be stored in the columns. The schema also defines relationships between tables and can include constraints to enforce data integrity, indexes to improve query performance, and other database-specific features such as triggers or stored procedures.

The main components of a database schema are as follows:

- **Tables**: Tables are collections of related data organized into rows and columns. Tables represent entities within the system, such as users or chats.

- **Columns**: Column attributes or fields of a table hold the data. Each column has a specific data type.

- **Data types**: Data types specify the kind of data that can be stored in a column, such as integers, text, dates, or Booleans.

- **Primary keys**: Primary keys are unique identifiers for table rows, ensuring that each record can be uniquely identified.

- **Foreign keys**: Foreign keys establish relationships between tables, linking rows of one table to rows of another.

- **Indexes**: Indexes boost the speed of fetching data from a database table but require more write operations and disk space to keep the index structure updated.

- **Constraints**: Constraints are rules applied to table columns to enforce data integrity, such as NOT NULL, UNIQUE, CHECK, or foreign key constraints.

Let's show the schema of our database for handling chat applications and discuss it.

Defining the database schema

We will define tables for users, chats, and messages, and relationships between the tables. We will specify the fields for every table, their names, types, and constraints. Then, we will define the relationships between the tables using foreign keys.

The CREATE TABLE command is used to create a new table in the database. Each CREATE TABLE statement defines the structure of the table by specifying its columns, their data types, and any constraints on those columns:

sql/schema.sql

```
CREATE TABLE "user"
(
```

Now, let's specify the columns.

SERIAL is a PostgreSQL data type used for auto-incrementing integer columns. It's commonly used for primary keys:

```
    id          SERIAL PRIMARY KEY,
```

Each new row inserted into the table without a specified value for this column will automatically get a unique integer value.

The PRIMARY KEY constraint uniquely identifies each row in a table. The SERIAL column is often used as a primary key.

Next, each column in the table is defined by its name followed by its data type and possibly one or more constraints, as follows:

```
    "createdAt" TIMESTAMP(3) DEFAULT CURRENT_TIMESTAMP NOT NULL,
    "updatedAt" TIMESTAMP(3) DEFAULT CURRENT_TIMESTAMP NOT NULL,
    name        VARCHAR(500)                           NOT NULL,
    email       VARCHAR(200)                           NOT NULL,
    password    VARCHAR(500)                           NOT NULL
);
```

Here are the constraints we see in the preceding code block:

- TIMESTAMP(3): This data type stores a date and time, with precision up to milliseconds (hence (3)). It's used for columns that record times, such as createdAt and updatedAt, which serve as indicators when the row was added to the database and changed the last time, respectively. These two fields are very useful for debugging purposes.

- VARCHAR (n): A variable character string type that can store up to n characters. For example, VARCHAR (500) can store strings up to 500 characters long.

- NOT NULL: This constraint ensures that a column cannot have a NULL value, meaning each row must have a value for this column.

- DEFAULT: This keyword is followed by a value or function that the column will use if no value is specified during an insert. For example, CURRENT_TIMESTAMP automatically stores the current date and time.

Now, we will create an index:

```
CREATE UNIQUE INDEX "user_email_key"
    ON "user" (email);
```

The CREATE UNIQUE INDEX command creates a unique index on the specified column(s) of a table. A unique index ensures that two rows cannot have the same value in these columns. For example, the email column in the user table cannot have duplicate values, enforcing unique email addresses for each user.

Next, we can define a chat table:

```
CREATE TABLE "chat"
(
    id          SERIAL PRIMARY KEY,
    "createdAt" TIMESTAMP(3) DEFAULT CURRENT_TIMESTAMP NOT NULL,
    "updatedAt" TIMESTAMP(3) DEFAULT CURRENT_TIMESTAMP NOT NULL,
    "ownerId"   INT                                    NOT NULL
REFERENCES "user"
        ON UPDATE CASCADE ON DELETE CASCADE,
    name        VARCHAR(1000)                          NOT NULL
);
```

We can see the following in the preceding code block:

- INT: Stands for integer, a numeric data type that can store whole numbers.

- REFERENCES: This sets up a foreign key relationship between two tables. It ensures that the value in this column must exist as a value in the referenced primary key of another table. For example, ownerId in the chat table references id in the user table.

- **Foreign key constraints with actions**: The REFERENCES keyword not only establishes a relationship between two tables but also supports defining actions upon updates or deletions:

  - ON UPDATE CASCADE: If the referenced entity is updated, the change is cascaded to the referring entity. For example, if a user's id property changes (which is rare and generally not recommended), all related chat records, which means their ownerId property that has the value of the user's id property, will be updated accordingly.

- ON DELETE CASCADE: If the referenced entity is deleted, all referring entities will also be deleted. For example, deleting a chat will delete all related message records.

Now, we will define a `message` table:

```
CREATE TABLE "message"
(
    id          SERIAL PRIMARY KEY,
    "createdAt" TIMESTAMP(3) DEFAULT CURRENT_TIMESTAMP NOT NULL,
    "updatedAt" TIMESTAMP(3) DEFAULT CURRENT_TIMESTAMP NOT NULL,
    "chatId"    INT                                    NOT NULL
REFERENCES "chat"
        ON UPDATE CASCADE ON DELETE CASCADE,
    type        VARCHAR(100)                           NOT NULL,
    message     TEXT                                   NOT NULL
);
```

The TEXT data type is used for long text strings. There's no limit to the length of text it can store.

Now that we know the schema, we can create it in the database.

Creating the database schema

First, we will need to connect to our database using the `psql` tool, which is a terminal-based frontend to PostgreSQL. To do that, we will need to identify the ID of our Docker PostgreSQL container by running the following command in a new terminal window:

```
$ docker ps
```

> **Important note**
> Be sure to not turn off the `postgres` container we started before.

You will see an output similar to this:

```
CONTAINER ID    IMAGE                           COMMAND
                    CREATED         STATUS
PORTS                                           NAMES
40ddbde6ae09    postgres:latest                 "docker-
entrypoint.s…"    15 minutes ago   Up 15
minutes                     0.0.0.0:5434->5432/
tcp                         chat_backend_2_postgres_1
```

The first value you see is the container ID; in my case, it is 40ddbde6ae09. Now, we can put the container ID in the following command to connect to psql in our Docker container:

```
$ docker exec -it 40ddbde6ae09 psql -U test test
```

Let's discuss what we see here:

- docker exec: Executes a command in a running container.
- -it: Combines the -i and -t flags. -i keeps **STDIN** (standard input) open even if not attached, and -t allocates a pseudo-**TTY** (teletype, which simulates a terminal), making the terminal interactive.
- 40ddbde6ae09: The container ID or name where the command is to be executed.
- psql: The command to be run inside the container, which is the PostgreSQL command-line interface.
- -U test: Specifies the username (test) to connect to the PostgreSQL server.
- test: The name of the database to connect to.

You will see that you entered the psql shell by seeing this line:

```
test=#
```

Now, you need to pass the SQL schema we discussed earlier in this section and press *Enter*. You should see the following output:

```
CREATE TABLE
CREATE INDEX
CREATE TABLE
CREATE TABLE
```

With these commands, we have managed to create our database and the schema of the tables that we will need for our chat application. Now, let's talk about how we can add and retrieve data from our database.

Writing CRUD SQL operations

Interacting with PostgreSQL involves using SQL, which is the standard language for relational database management and manipulation. It allows us to perform various operations on the data stored within our database, structured around the fundamental concepts of CRUD.

Let's discuss the most command SQL command types and their purposes:

- **Data definition language**: Commands that define the database structure. Examples include CREATE, ALTER, and DROP, which can be used to create, modify, and delete database objects such as tables and indexes.

- **Data manipulation language**: Commands that handle data manipulation within the tables. These include INSERT (Create), SELECT (Read), UPDATE (Update), and DELETE (Delete).

- **Data control language**: Commands that deal with rights, permissions, and other controls of the database system. Examples are GRANT and REVOKE.

- **Transaction control language**: Commands that manage transactions in the database, such as COMMIT and ROLLBACK.

Now, let's play around from inside our psql tool with our database – we will create a new user, associate a chat and message with the user, then retrieve messages and chats, and finally update and remove the user:

1. **Inserting a new user**:

```
INSERT INTO "user" (name, email, password)
VALUES ('Jane Doe', 'jane.doe@example.com', 'securePassword');
```

This statement adds a new row to the user table with values for the name, email, and password columns. Each value corresponds to the column specified in the INSERT INTO clause. This operation creates a new user record in the database.

2. **Starting a new chat**:

```
INSERT INTO "chat" ("ownerId", name)
VALUES (1, 'LLM fun');
```

Here, we insert a new row into the chat table. The ownerId column is set to 1, assuming this is the ID of Jane Doe from the previous insert. The chat is named LLM fun.

3. **Creating a new message in the chat**:

```
INSERT INTO "message" ("chatId", type, message)
VALUES (1, 'text', 'Welcome to my tech corner!');
```

This command inserts a new message into the message table. chatId is set to 1, linking this message to the previously created chat. The type column specifies the message format (in this case, 'text'), and the message column contains the actual message content.

4. **Retrieving all users**:

```
SELECT * FROM "user";
```

This query fetches all columns (*) for every row in the user table, showing a complete list of users.

5. **Retrieving all messages from a specific chat:**

```
SELECT message FROM "message" WHERE "chatId" = 1;
```

This SELECT statement retrieves all messages belonging to the chat with chatId equal to 1. It filters rows using the WHERE clause to match the given condition.

6. **Updating a user's email:**

```
UPDATE "user" SET email = 'new.jane.doe@example.com' WHERE id =
1;
```

This command updates the email column of the user with id = 1. The SET clause specifies the new value for the column to be updated, and the WHERE clause ensures that only the record for the specified user ID is updated.

7. **Removing a chat:**

```
DELETE FROM "chat" WHERE id = 1;
```

The DELETE FROM statement removes rows from the chat table where the condition in the WHERE clause (id = 1) is met. This deletes the chat.

> **Important note**
> The DELETE FROM operation will cascade and delete any related messages due to the foreign key constraint with ON DELETE CASCADE.

Summary

In this chapter, we started a critical journey into persistent storage, focusing on the integration of PostgreSQL within Docker for our chat application. We covered the essentials of Docker, how to set up PostgreSQL as a Docker container, the creation of a comprehensive database schema, and the fundamentals of CRUD operations. This step is crucial for implementing the dynamic functionalities of our chat application, enabling real-time data processing and manipulation, and further enhancing our backend development skills for building sophisticated, data-driven web applications.

The following two chapters will take us deeper into the practical aspects of application development by teaching us how to interact with our PostgreSQL database using SQL directly from our application. We will use raw SQL queries to interact with our database in *Chapter 8* and **object-relational mapping (ORM)** in *Chapter 9*.

Interacting with PostgreSQL Using Libraries

In the previous chapter, we established a foundation for persistent storage by setting up PostgreSQL within Docker, enabling us to perform basic SQL queries. This setup is crucial for storing data persistently across server sessions, directly impacting our application's reliability and functionality.

Now, we'll take a step forward by integrating SQL operations into our server. This chapter will focus on utilizing the pg library to interact with our PostgreSQL database from within our server code. You'll learn how to seamlessly connect your backend server with the database, perform data operations, and ensure that your data handling is both efficient and secure.

The following is what we'll cover in this chapter:

- Integrating SQL implementation in the codebase
- Testing our SQL integrations to ensure reliability

This knowledge bridges the gap between your application and its data layer, enabling you to build dynamic, data-driven features with confidence. By the end of this chapter, you'll be equipped to handle data manipulation directly from your backend, enhancing your application's overall performance and capabilities.

Technical requirements

We need to install pg. We can do this with the following command:

```
$ bun add pg@^8.11.3
```

We also need to install pg types as a dev dependency so that the library provides types for our TypeScript:

```
$ bun add @types/pg@^8.10.9 -d
```

All the code we are going to discuss in this chapter is available at `https://github.com/PacktPublishing/Full-Stack-Web-Development-with-TypeScript-5/tree/main/Chapter08`.

Integrating SQL implementation in the codebase

In *Chapter 4*, we created an interface for our database interactions and integrated it throughout our codebase. To switch to the new SQL implementation, we need to create implementations of the `IDatabaseResource` interface for users, chats, and messages. Starting with users, we'll write a class that conforms to `IDatabaseResource<DBUser, DBCreateUser>` to handle user management in our database.

Writing a class

In the following code block, we define `UserSQLResource` to handle our users using SQL. We will implement the `IDatabaseResource` interface, which contains quite a few methods, so we will explain each function after the other:

src/storage/sql.ts

```
import type { Pool } from "pg";
import type {
  DBChat,
  DBCreateChat,
  DBCreateMessage,
  DBCreateUser,
  DBMessage,
  DBUser,
} from "../models/db";
import type { IDatabaseResource } from "./types";

export class UserSQLResource
  implements IDatabaseResource<DBUser, DBCreateUser>
{
  pool: Pool;

  constructor(pool: Pool) {
    this.pool = pool;
  }
```

Here are the initialization and constructor as seen in the preceding code block:

- `pool: Pool;`: Declares a class property named `pool` of type `Pool` from the pg library. This `Pool` object is used to manage connections to the PostgreSQL database. `Pool` helps us to reuse the connection established to the database.

- `constructor(pool: Pool) { this.pool = pool; }`: The constructor takes a `Pool` object as an argument and assigns it to the class's `pool` property. This `Pool` instance is then used for executing SQL queries.

Let's implement the `create` method now:

```
async create(data: DBCreateUser): Promise<DBUser> {
  const query =
    'INSERT INTO "user" (name, email, password) VALUES ($1, $2, $3)
RETURNING *';
  const values = [data.name, data.email, data.password];
  const result = await this.pool.query(query, values);
  return result.rows[0] as DBUser;
}
```

Here are the details of the preceding code lines:

- Placeholder parameters: `$1`, `$2`, `$3` are parameter placeholders used in the SQL queries. These are replaced by the values provided in the `values` array when the query is executed. This approach helps prevent SQL injection attacks by separating the query structure from the data it operates on, so that no malicious values can be inserted from the user input.

- Query execution: `await this.pool.query(query, values)` executes the SQL query through the PostgreSQL connection pool. This asynchronous operation waits for the query to complete and returns the result. Running I/O operations can be expensive – by making them asynchronous, we ensure that our code can make use of idle time, allowing the CPU to do other tasks while waiting for the I/O operation to finish. This helps optimize the use of our computer's resources.

- Processing the result: `result.rows[0] as DBUser` accesses the first row of the result set and casts it to the `DBUser` type. This is used when expecting a single record in response (common in `create`, `delete`, and `get` operations).

- The `create` method: The `'INSERT INTO "user" (name, email, password) VALUES ($1, $2, $3) RETURNING *';` line is a SQL command that inserts a new record into the `"user"` table with values for the name, email, and password. The method returns a Promise that resolves with the result of the query, which is the created user.

Now let's turn to the `delete` method:

```
async delete(id: string): Promise<DBUser | null> {
  const query = 'DELETE FROM "user" WHERE id = $1 RETURNING *';
  const values = [id];
  const result = await this.pool.query(query, values);
  return result.rowCount ?? 0 > 0 ? (result.rows[0] as DBUser) :
null;
}
```

The `'DELETE FROM "user" WHERE id = $1 RETURNING *';` SQL query deletes a record from the `"user"` table where the `id` value matches the provided value. The `RETURNING *` part returns the deleted row(s), allowing the method to provide feedback about the operation's result.

`result.rowCount ?? 0 > 0` checks whether any rows were affected (or returned) by the query. This is useful for operations such as `delete`, where the existence of affected rows indicates success.

Now let's write the `get` method:

```
async get(id: string): Promise<DBUser | null> {
  const query = 'SELECT * FROM "user" WHERE id = $1';
  const values = [id];
  const result = await this.pool.query(query, values);
  return result.rowCount ?? 0 > 0 ? (result.rows[0] as DBUser) :
null;
}
```

The `'SELECT * FROM "user" WHERE id = $1'` SQL query selects all columns from the `"user"` table for the row(s) where the `id` value matches the provided value. This retrieves the user's details based on the provided `id`.

Let's define the `find` and `findAll` functions, which are going to use the same `findByFields` private function with a param for returning one or multiple results:

```
async find(data: Partial<DBUser>): Promise<DBUser | null> {
  return this.findByFields(data, false);
}

async findAll(data: Partial<DBUser>): Promise<DBUser[]> {
  return this.findByFields(data, true);
}
```

Now let's define the `findByFields` function, which is going to iterate the fields in the passed data and create a search query based on that:

```
private async findByFields<T extends (DBUser | null) | DBUser[]>(
```

The generic type <T> method is a generic function that returns a promise of type T, which can be either a single DBUser object, null, or an array of DBUser objects. The return type is determined by the all parameter.

Let's continue with defining the parameters for the method:

```
    data: Partial<DBUser>,
    all: boolean = false,
  ): Promise<T> {
    const fields: string[] = [];
    const values: unknown[] = [];

    Object.keys(data).forEach((key, index) => {
      fields.push(`"${key}" = $${index + 1}`);
      values.push(data[key as keyof DBUser]);
    });

    const whereClause =
      fields.length > 0 ? `WHERE ${fields.join(" AND ")}` : "";
    const query = `SELECT *
                        FROM \"user\" ${whereClause}`;
```

Here's what is happening in the preceding code block:

- A dynamically constructed SQL query: The method constructs a WHERE clause dynamically based on the keys and values of the data parameter, which is a partial representation of DBUser. This allows for flexible queries based on the provided fields.

- A parameterized query with safe placeholder values: Similar to previous methods, it uses parameter placeholders ($1, $2, etc.) to prevent SQL injection. The placeholders are replaced by the corresponding values in the values array.

Here is the conditional return based on the all parameter:

```
  const result = await this.pool.query(query, values);
      return all
        ? (result.rows as T)
        : result.rowCount ?? 0 > 0
          ? (result.rows[0] as T)
          : (null as T);

  }
```

If `all` is true, the method returns all matching rows as `DBUser[]`. If false, it returns the first matching row as `DBUser` or `null` if no match is found.

Next, let's cover the last method to update a user.

Updating a user

Now, we will define a function to update the user:

```
async update(id: string, data: Partial<DBUser>): Promise<DBUser |
null> {
    const fields: string[] = [];
    const values = [];

    Object.keys(data).forEach((key, index) => {
      fields.push(`"${key}" = $${index + 1}`);
      values.push(data[key as keyof DBUser]);
    });

    values.push(id); // Push the id as the last parameter
    const setClause = fields.join(", ");
    const query = `UPDATE \"user\"
                    SET ${setClause}
                    WHERE id = $${fields.length + 1} RETURNING *`;

    const result = await this.pool.query(query, values);
    return result.rowCount ?? 0 > 0 ? (result.rows[0] as DBUser) :
null;
  }
}
```

Let's discuss what the `update` function does:

- Updates user records: This method updates user records in the database based on the provided `id` and partial user data (`data`). It constructs a dynamic `SET` clause to update only the specified fields.

- Dynamic `SET` clause construction: Similar to `findByFields`, `update` constructs the `SET` clause dynamically based on the keys and values in the data object. Each field to be updated is added to the fields array with its corresponding placeholder for value substitution.

- Appending `id` to values: The `id` of the record to be updated is appended to the end of the values array. This `id` is used in the `WHERE` clause to identify the record to be updated.

- Returning the updated record: The `RETURNING *` part of the SQL query ensures that the updated record is returned by the query.

With this, we have written a working class that uses SQL to interact with our user tables. We also need to write similar classes to interact with chats and messages. They are very similar, so I'm not going to explain the code here, but you can go through the code for all the classes and copy it from here:

`https://github.com/PacktPublishing/Hands-On-Full-Stack-Web-Development-with-TypeScript-5/blob/main/Chapter08/chat_backend/src/storage/sql.ts`

If the class implementations look repetitive to you – they are. They indeed reuse a lot of similar SQL queries and code logic. We are going to solve this problem in the next chapter with the use of ORM and a single class to interact with the database.

Now, we can integrate our SQL implementation into our main function.

Incorporating SQL implementation into the main function

Add the `import` lines after all the other imports and add the new function to the end of the file:

src/controllers/main.ts

```
...
import { Pool } from "pg";
import {ChatSQLResource, MessageSQLResource, UserSQLResource} from
"../storage/sql";


...
export function createSQLApp() {
  const pool = new Pool({
    connectionString: Bun.env.DATABASE_URL,
  });
  return createMainApp(
    createAuthApp(new UserSQLResource(pool)),
    createChatApp(new ChatSQLResource(pool), new
MessageSQLResource(pool)),
  );
}
```

The preceding code initiates a database connection pool from an environment variable, then sets up `auth` and `chat` apps with the database pool we have established.

Now, instead of creating our app with in-memory storage, we can swap it for the SQL implementation.

Replacing in-memory data storage with a SQL-based solution

Let's now change our entrypoint file to use a SQL-based app:

src/index.ts

```
import { createSQLApp } from "./controllers/main";
const app = createSQLApp();
export default app;
```

To get our database up and running, we simply need to start the Docker container and ensure the DATABASE_URL is correctly set in our environment variables.

From the previous chapter, we have a Docker Compose file ready for both testing and development. Now, we'll enhance it by adding an extra volume that loads the SQL schema when the container starts. Just include these lines in the volumes section of both Docker Compose files for the development containers:

docker_compose.yml

```
...
    volumes:
      - ./data/postgres:/var/lib/postgresql/data
      - ./data/init_scripts:/docker-entrypoint-initdb.d
```

Now, include these lines in the volumes section of both Docker Compose files for the testing containers:

docker_compose_test.yml

```
...
    volumes:
      - ./data/init_scripts:/docker-entrypoint-initdb.d
```

We've introduced a new script into the docker-entrypoint-initdb.d directory, ensuring automatic execution upon container startup. Importantly, the schema.sql file from the previous chapter should be placed in the ./data/init_scripts folder. Since our test container lacks a persistent volume, each restart clears the database, providing a fresh schema setup—ideal for testing scenarios.

When you start the containers, it will also execute our SQL file and will create the schema for us. To activate the container, use this command:

```
$ docker-compose -f docker_compose.yml up -d
```

With the database operational, integrate the connection string into your .env.dev file as follows:

```
DATABASE_URL="postgresql://user:pass@localhost:5433/db"
```

It's crucial to manage database connections in separate environment files for development and testing. This is because Bun defaults to loading the .env file, overriding variables in any subsequently loaded environment files.

If we launch the app now, we will read and write data to the database instead of the in-memory storage we used before, ensuring data persistence across app restarts:

```
$ bun run dev --env-file .env.dev
```

Congrats! We have successfully replaced our in-memory data storage with a SQL-based solution, which is very common to use in a production setup.

> **Important note**
>
> Remember that you can always clear the data created in your database and start again. For this, you need to stop the Docker containers you are running and remove the data directory for PostgreSQL:
>
> ```
> $ docker-compose -f docker_compose.yml down
> ```
>
> ```
> $ rm  -r data/postgres
> ```
>
> And don't forget that after this, you need to recreate the required tables in the database, as shown in the previous chapter – the only thing different is the user and db values we need to use to connect to the database:
>
> ```
> $ docker exec -it $DOCKER_PG_CONTAINER_ID psql -U user db
> ```

Let's talk now about what we need to do to test our SQL integrations.

Testing our SQL integrations to ensure reliability

We have created the implementation using SQL, but we haven't tested whether it works with our new implementation. Fortunately, to complete that, we do not need to do much. Our test's setup is flexible enough to run the test with any data source we need, but we need to ensure that we provide the data source correctly and that we clean it between the test runs. This is what we are going to accomplish with the following code snippets:

1. First, start the test Docker container:

    ```
    $ docker-compose -f docker_compose_test.yml up -d
    ```

2. Then, we will add utils for our test (`tests/utils.ts`) that are going to help us clean the database between each test run:

    ```
    import type { Pool } from "pg";

    export async function resetSQLDB(pg: Pool) {
      await pg.query(
        `
        DELETE FROM message;
        DELETE FROM chat;
        DELETE FROM "user";`,
      );
    }
    ```

    Here, `resetSQLDB` deletes all the data from all the tables we use.

3. We also need to change the setup of our test files (`tests/auth.test.ts`) to use the SQL-based storage when they run instead of the in-memory storage:

    ```
    import { beforeEach, describe, expect, test } from "bun:test";
    import { Pool } from "pg";
    import { createSQLApp } from "../src/controllers/main";
    import { resetSQLDB } from "./utils";

    describe("auth tests", () => {
      const app = createSQLApp();

      const pool = new Pool({
        connectionString: Bun.env.TEST_DATABASE_URL,
      });
    ```

```
    beforeEach(async () => {
      await resetSQLDB(pool);
    });

    test("POST /register - normal case", async () => {
    ....
```

This setup is very similar to what we wrote in *Chapter 5*. The only difference is that as we store the data in permanent storage, we cannot simply recreate the app between the test runs and start with clear storage. Now, we run our `resetSQLDB` utility function to clean the database before every test run. In a similar manner, we need to change the setup in our `tests/chat.test.ts` file.

4. Before we run the tests, we also need to provide the environment variable with the connection string to our database. Add the following to `.env.test`:

    ```
    DATABASE_URL="postgresql://test:test@localhost:5434/test"
    ```

5. Now, use the test environment variables and run the tests with this command:

    ```
    $ bun test -env-file .env.test
    ```

    Our tests use the test database we started with Docker for running the tests, and they also clear the data between the runs, so we always start with a clean database.

This is everything that we need to integrate with SQL in all parts of our application.

Summary

In this chapter, we've taken a significant step by integrating our backend server with PostgreSQL, utilizing SQL and the pg library for direct database interactions. This integration allows us to efficiently store, retrieve, and manipulate data within our server, laying the groundwork for a robust and dynamic backend system. This knowledge is crucial for developing advanced backend functionalities, enabling us to manage application data effectively.

Moving forward, in the next chapter, we will explore the use of **Object-Relational Mapping** (**ORM**) to interact with our database. ORM will help us reduce code repetition and simplify database interactions, making our development process more efficient and our codebase cleaner. This next phase is important for enhancing our application's maintainability and scalability.

9

Interacting with PostgreSQL Using Prisma ORM

In the previous chapter, we delved into the world of persistent storage, focusing on the implementation and management of databases within Docker containers. We learned how to set up PostgreSQL in Docker, construct a robust database schema for a chat application, and perform essential CRUD operations directly with SQL. This knowledge is pivotal in ensuring data persistence and integrity, critical for maintaining user information and messages across server restarts in dynamic applications.

This chapter shifts our focus toward streamlining database interactions and migrations in web development. We introduce **object-relational mapping (ORM)** as a powerful tool to abstract and simplify database operations, specifically through **Prisma**, a next-generation ORM. By integrating Prisma into our development workflow, we aim to enhance productivity, reduce errors, and promote code clarity when interacting with our PostgreSQL database.

Understanding and utilizing ORMs, especially Prisma, is invaluable for developers. It not only simplifies CRUD operations with its intuitive query language but also facilitates database schema migrations, making it easier to evolve your database alongside your application without the hassle of manual SQL script management.

The topics we'll cover in this chapter are the following:

- Introduction to ORMs and Prisma
- Handling migrations using Prisma
- Interacting with the database using Prisma
- Testing our ORM integration:

Starting with Introduction to ORMs and Prisma, we'll break down the concept of ORMs and discuss the advantages of using Prisma over traditional SQL queries. This section will set the stage for understanding how Prisma can make our development process more efficient and our code base cleaner and more maintainable.

Technical requirements

To proceed with this chapter, you need to install `prisma` and `@prisma/client` to use Prisma to interact with our database. You can do it with this command:

```
$ bun add prisma@^5.7.1 @prisma/client@^5.7.1
```

Another tool that we need to install is `dotenv`. This tool will help us to load environment variables from files before the Bun execution. We can install it with the following command globally for our system:

```
$ npm install -g dotenv-cli@^7.4.2
```

All the code we are going to discuss in this chapter is available at `https://github.com/PacktPublishing/Full-Stack-Web-Development-with-TypeScript-5/tree/main/Chapter09`.

Introduction to ORMs and Prisma

At the heart of modern application development, ORM serves as a crucial bridge between the objects in our TypeScript applications and the tables in a relational database. This technique allows us to interact with our database using the programming language we're already working with, sidestepping the direct use of SQL. Essentially, ORMs let us treat database records as objects in our code, simplifying data manipulation and management.

Here is why ORMs shine over plain SQL:

- **Boosted productivity**: By automating the creation of SQL queries, ORMs free us from manual query writing, allowing us to focus on building features
- **Cleaner code**: They encapsulate the boilerplate of database interactions, making our code base more readable and maintainable
- **Ease of maintenance**: With ORMs, adapting to schema changes becomes more straightforward, ensuring our application evolves smoothly
- **Flexibility across databases**: Many ORMs support multiple database systems, making it easier to switch or integrate different databases without changing our application logic

However, it's not all smooth sailing. Here are some potential disadvantages of using ORMs:

- **Performance overhead**: The abstraction layer can sometimes generate less optimized queries than hand-crafted SQL, affecting performance
- **Complexity for simple queries**: For straightforward database operations, ORMs can introduce unnecessary complexity

- **Steep learning curve**: Fully leveraging an ORM's capabilities often requires a deep understanding of its features and behaviors

- **Loss of control**: Advanced database optimizations may be harder to implement due to the abstraction over SQL

All in all, ORMs are still very useful tools for applications with a lot of easy-to-medium-complexity queries. So, let's discuss the ORM that we are going to use in our project.

Introducing Prisma – our ORM of choice

Prisma emerges as an ORM tailored for TypeScript, offering a compelling suite of features designed to streamline database operations. What sets Prisma apart is its focus on type safety and developer experience, ensuring that we write secure and efficient database interactions with minimal effort. Here is what Prisma brings to the table:

- **Type safety**: Prisma's approach to type safety is unparalleled, providing us with autogenerated types for our models. This integration drastically reduces the risk of runtime errors, making our database interactions predictable and robust.

- **Developer experience**: With intelligent autocompletion and a comprehensive query builder, Prisma turns database operations into a more seamless experience, enhancing productivity and reducing bugs.

- **Migration management**: Prisma Migrate addresses the complexities of database schema evolution, offering a streamlined process for applying and tracking schema changes.

Prisma also comes with a solution for the management of changes to our SQL schema, which is called **Prisma Migrate**. Let's talk more about it.

Understanding Prisma Migrate

Prisma Migrate is a feature within Prisma that automates the process of evolving your database schema in a safe, easy, and version-controlled manner. It helps in creating, applying, and managing database schema changes through simple commands. This feature tracks schema changes in versioned files, allowing for collaborative development and straightforward deployment processes.

When you modify your Prisma schema file, Prisma Migrate can generate a corresponding SQL migration script that reflects the changes. These scripts are applied to your database to update its schema, ensuring that your database and application model stay in sync.

With this, we are ready to look into how we are going to handle migrations in our system using Prisma.

Handling migrations using Prisma

To handle migrations and schema using Prisma, we need to define our database schema using the Prisma format so that it can understand which SQL operations it transforms to.

We will first define the database provider we are going to use, and then we will define the structure for each table. As a result, we will get migrations that we can run to generate a database schema and also classes for interaction with the database from inside our code.

Defining the database schema

Let's now write our `user` table in Prisma format:

prisma/schema.prisma

```
generator client {
  provider = "prisma-client-js"
}
```

These lines specify that a Prisma client should be generated for JavaScript/TypeScript. It means the Prisma client is going to create classes in our code that we can call within our TypeScript application to interact with the database.

In the following code lines, `provider` indicates the type of database. In this case, it's PostgreSQL, much like specifying the database engine in a connection string in SQL:

```
datasource db {
  provider = "postgresql"
```

The following code line defines the database connection string through an environment variable:

```
  url       = env("DATABASE_URL")
}
```

Now, we will create a `User` model that represents the `user` table:

```
model User {
```

Here is an `id` integer that serves as the primary key, auto-incrementing for each new record. This mirrors the `SERIAL` primary key in SQL:

```
  id        Int      @id @default(autoincrement())
```

Next, we have timestamps for record creation and last update, automatically managed by Prisma:

```
createdAt DateTime @default(now())
updatedAt DateTime @updatedAt
```

These are the user attributes, with the `email` field marked as unique:

```
name     String    @db.VarChar(500)
email    String    @unique @db.VarChar(200)
password String    @db.VarChar(500)
```

The `@db.VarChar(n)` annotations specify the string length, directly mapping to SQL's VARCHAR(n) type definition.

Here is a one-to-many relationship with the `Chat` model, indicating that a user can have multiple chats. In SQL, we managed this relationship with a `user_id` foreign key in the `chat` table:

```
chats     Chat[]

}
```

Next, let's write our `chat` table in Prisma format:

```
model Chat {
    id         String        @id @default(uuid())
    createdAt DateTime @default(now())
    updatedAt DateTime @updatedAt
    ownerId   String
```

An integer field represents a foreign key link to the `User` model. In SQL, this would be a column with a foreign key constraint pointing to the `user` table.

The following lines establish a many-to-one relationship to `User`, connecting each chat to its owner. This is a declarative way to handle foreign key relationships in Prisma:

```
name     String    @db.VarChar(1000)
owner    User      @relation(fields: [ownerId], references: [id])
```

The following stores the foreign key to the `Chat` model, linking each message to a chat:

```
messages  Message[]
}
```

Next, let's define our `message` table:

```
model Message {
    id         String      @id @default(uuid())
    createdAt DateTime @default(now())
    updatedAt DateTime @updatedAt
    chatId     String
```

The following lines establish a foreign key relationship to `Chat`, ensuring each message is associated with a specific chat:

```
    type       String     @db.VarChar(100)
    message    String     @db.Text
    chat       Chat       @relation(fields: [chatId], references: [id])
}
```

Now, that we've created our schema, let's apply the migration to our database.

Applying the migration to our database

To apply the migration to our database, we will follow these steps:

1. First, let's stop our `dev` database container if it is running:

   ```
   $ docker-compose -f docker_compose.yml down
   ```

2. Then, we should delete the `data/postgres` folder as we will need to instantiate the database again.

3. We also need to remove the `data/init_scripts` volume from both our `docker_compose.yml` and `docker_compose_text.yml` files as we no longer want to create the schema using native SQL when we start the container.

4. Now, run the container again:

   ```
   $ docker-compose -f docker_compose.yml up -d
   ```

5. Now, we can run `prisma` to create and apply migrations to our database with the following command:

   ```
   $ bun --env-file=.env.dev run prisma migrate dev --name init
   ```

> **Important note**
>
> If you encounter an issue on Windows that goes along the lines of `Environment variable is not found postgresql://user:pass@localhost:5433/db`, you need to either use a Bash-compatible terminal such as Git Bash, use WSL, or export the environment variable manually before you run the command. You can export the variable manually by executing `$ set DATABASE_URL=postgresql://user:pass@localhost:5433/db`

6. You are going to see the following message in the terminal:

    ```
    ? We need to reset the "public" schema at "localhost:5433"
    Do you want to continue? All data will be lost. > (y/N)
    ```

    Type y and press *Enter*.

    You should see the following message:

    ```
    Your database is now in sync with your schema.
    ```

    Also, you will notice that there is a new folder under `prisma` called `migrations`. The `migrations` folder keeps track of changes that we applied to the database. When we change our database, we will create a new migration, which we are going to apply. This guarantees that we always keep track of changes we make to our database and can reliably recreate the database structure in any environment.

    In production, you cannot just remove the database and start again; if you make a change to the schema, you need to apply the changes. This is exactly what migrations help us achieve.

With this in place, we can move to the next part of this chapter – how to interact with our database from our code using Prisma.

Interacting with the database using Prisma

Now that we have the database ready for interactions, as well as our interaction client generated with the necessary classes, we can write an implementation of our `IDatabaseResource` interface using Prisma. First, we are going to import the `prisma` client that contains objects we will use to interact with our tables; then we are going to implement CRUD methods delegating the main work to `prisma`.

Defining the Prisma Client class

Let's define a class that is going to implement `IDatabaseResource` and use `prisma` for database interaction.

src/storage/orm.ts

The following line imports `PrismaClient` for database interaction:

```
import type { PrismaClient } from "@prisma/client";
```

The following code block initializes the class with a `PrismaClient` instance:

```
import type {
  DBChat,
  DBCreateChat,
  DBCreateMessage,
  DBCreateUser,
  DBMessage,
  DBUser,
} from "../models/db";
import type { IDatabaseResource } from "./types";

export class UserDBResource implements IDatabaseResource<DBUser,
DBCreateUser> {
  prisma: PrismaClient;

  constructor(prisma: PrismaClient) {
    this.prisma = prisma;
  }
```

The following code block uses Prisma client's `create` method on the `User` model, spreading the `data` object to match the `create` method's expected input:

```
async create(data: DBCreateUser): Promise<DBUser> {
  const user = await this.prisma.user.create({
    data: { ...data },
  });
  return user as DBUser;
}
```

The following code lines utilize Prisma client's `delete` method, specifying which user to delete by using a `where` clause with the user's ID:

```
async delete(id: string): Promise<DBUser | null> {
  const user = await this.prisma.user.delete({ where: { id: id } });
  return user as DBUser;
}
```

The following code snippet uses Prisma client's `findFirst` method to search for the first user matching the given ID. A `where` clause is used to specify the search condition:

```
async get(id: string): Promise<DBUser | null> {
  const user = await this.prisma.user.findFirst({ where: { id: id }
});
  return user as DBUser | null;
}
```

Let's continue with the functions that are left in `UserDBResource` and define the `find`, `findAll`, and `update` methods.

These lines employ the `findFirst` method to locate the first user record that fits the search criteria provided in `data`:

```
async find(data: Partial<DBUser>): Promise<DBUser | null> {
  const user = await this.prisma.user.findFirst({ where: { ...data }
});
```

The following code lines use the `findMany` method to fetch multiple user records that match the conditions in `data`:

```
  return user as DBUser | null;
}

async findAll(data: Partial<DBUser>): Promise<DBUser[]> {
  const users = await this.prisma.user.findMany({ where: { ...data }
});
```

The following code snippet utilizes the `update` method, specifying the user to update through a `where` clause and applying the changes provided in `data`:

```
  return users as DBUser[];
}

async update(id: string, data: Partial<DBUser>): Promise<DBUser |
null> {
  const updateUser = await this.prisma.user.update({
    where: {
```

```
        id,
      },
      data,
    });
    return updateUser as DBUser | null;
  }
}
```

As you can see, our Prisma client generated very comfortable classes that we can use to interact with the database, which simplifies and abstracts the way we work with the database.

We also need to implement ChatDBResource and MessageDBResource, but as the code there is very similar, you can simply copy these classes from here:

https://github.com/PacktPublishing/Full-Stack-Web-Development-with-TypeScript-5/blob/main/Chapter09/chat_backend/src/storage/orm.ts

We now have the ORM implementation, and we can integrate it into our main function.

Integrating ORM into the main function

To do this, add import lines after all the other imports and add the new function to the end of the file:

src/controllers/main.ts

```
...
import { PrismaClient } from "@prisma/client";
import {
  ChatDBResource,
  MessageDBResource,
  UserDBResource,
} from "../storage/orm";



...
export function createORMApp() {
  const prisma = new PrismaClient();
  prisma.$connect();
  return createMainApp(
    createAuthApp(new UserDBResource(prisma)),
    createChatApp(new ChatDBResource(prisma), new
MessageDBResource(prisma)),
  );
}
```

The preceding code imports our Prisma client so that we can create it and then sets up `auth` and `chat` apps with the Prisma client. `prisma.$connect();` is used to explicitly initiate the connection to our Prisma client after we create a client. While it is not strictly necessary, it is still a good practice.

Now, instead of creating our app with SQL storage, we can swap it for the ORM implementation:

src/index.ts

```
import { createORMApp } from "./controllers/main";
const app = createORMApp();
export default app;
```

With this in place, we can launch the server, which is going to now use our ORM implementation to talk to the database:

```
$ bun --env-file=.env.dev dev
```

Nice! We have replaced the SQL-based solution with an ORM one, which has a lot of pros, especially for bigger applications.

The only thing left is to correctly interact with our database during testing, so let's cover that next.

Testing our ORM integration

We need to see that our tests pass with the newly created ORM integration and that everything works as expected. To do this, we will need to apply our migrations to the test Docker database and change how we instantiate the tests. To do so, follow these steps:

1. First, start the test Docker container:

    ```
    $ docker-compose -f docker_compose_test.yml up -d
    ```

2. Now, we can apply our migrations to the test database with the following command:

    ```
    $ bun --env-file=.env.dev run prisma migrate deploy
    ```

    First, we provide the test environment variables, and then we run all the migrations we have created for our test database, which creates all the tables.

3. Now, let's add an extra function to our test utils file (`tests/utils.ts`) to clean up between the tests using Prisma client:

    ```
    import type { PrismaClient } from "@prisma/client";
    import type { Pool } from "pg";

    export async function resetORMDB(prisma: PrismaClient) {
      await prisma.$transaction([
    ```

```
      prisma.message.deleteMany(),
      prisma.chat.deleteMany(),
      prisma.user.deleteMany(),
   ]);
}
...
```

Let's discuss the new `resetORMDB` function here. `resetORMDB` runs a transaction inside `prisma` that deletes all the information in all our tables using `deleteMany` without any query, which wipes out the whole table.

4. We also need to change the setup of our test files (`tests/auth.test.ts`) to use the ORM-based implementation when they run instead of the SQL-based implementation:

```
import { PrismaClient } from "@prisma/client";
import { beforeEach, describe, expect, test } from "bun:test";
import { createORMApp } from "../src/controllers/main";
import { resetORMDB } from "./utils";

describe("auth tests", () => {
  const app = createORMApp();
  const prisma = new PrismaClient();

  beforeEach(async () => {
    await resetORMDB(prisma);
  });

  test("POST /register - normal case", async () => {
....
```

This setup is very similar to what we wrote in *Chapter 5*. The only difference is that we instantiate our app using `createORMApp` and then run our `resetORMDB` utility function to clean the database before every test run. You will also need to make a similar setup in the `tests/chat.test.ts` file.

5. Now, you can run the tests with this command:

```
$ bunx dotenv -e .env -e .env.test -- bun test
```

This is everything that we need to integrate with the ORM in all parts of our application.

Summary

In this chapter, we delved into the integration of ORMs, with a focus on Prisma, into our development workflow. We started with an introduction to ORMs and Prisma, highlighting their significance in bridging the gap between databases and application logic. We then explored handling database schema changes using Prisma's migration tools, followed by learning to interact directly with the database using Prisma's client. Finally, we covered testing our ORM integration to ensure reliability and stability in our application's data layer.

This integration is crucial for enhancing our application's development efficiency and maintainability by abstracting complex database operations, making them more manageable and less error-prone. It also ensures that our application remains robust and adaptable to changes, thanks to streamlined schema migrations and thorough testing of data interactions.

In the next chapter, we shift our focus to integrating external APIs, specifically starting with the groundwork needed to integrate the OpenAI API. We'll start with the basics of integrating external APIs with TypeScript and Hono, a crucial step toward enriching our application's functionality by generating dynamic responses to user messages. This step is essential for developing interactive applications that leverage the power of external services.

Part 4:
AI Integration with OpenAI API

This part introduces the integration of **artificial intelligence** (**AI**) into your applications using the OpenAI API. It focuses on how to set up and utilize AI technologies to enhance application capabilities and provide advanced features such as natural language processing and machine learning. This part is particularly valuable for developers looking to integrate cutting-edge AI functionalities into their applications.

This part includes the following chapters:

- *Chapter 10, Basics of Integrating External APIs with TypeScript and Hono*
- *Chapter 11, Setting up and Configuring the OpenAI API for the Backend*

10

Basics of Integrating External APIs with TypeScript and Hono

In the previous chapter, we learned how to handle our database to store persistent data. Now, it's time to finish our backend server by adding an API integration with OpenAI. We will focus here on using `fetch` – a Promise-based method for making HTTP requests from our Bun environment – to communicate with external services. Then, we will see how to improve the reliability of our API integration by handling errors and introducing retries. Also, we will make sure that we get a response in the format we want with the help of response validation. All this knowledge is going to enable us to write robust and effective API integrations.

In this chapter, we're going to cover the following main topics:

- Introduction to API integration in TypeScript using `fetch`
- Handling errors and retries
- Validating API correctness

We will start by talking about the importance of calling external APIs and how to do this with `fetch`.

Technical requirements

To proceed with this chapter, you don't need to install any additional libraries.

All the code we are going to discuss in this chapter is available at `https://github.com/PacktPublishing/Full-Stack-Web-Development-with-TypeScript-5/tree/main/Chapter10`.

Let's get into our chapter now.

Introduction to API integration in TypeScript using fetch

Integrating external APIs is a cornerstone of building dynamic and feature-rich applications. We need to call external services to enrich the functionality of our application. For the sake of this book, we will need to integrate the OpenAI API to generate **Generated Pre-trained Transformer (GPT)** responses. Integrating with external APIs is essential but comes with a few things to watch out for: the API call can take a significant time, it can fail, and we can get back something we do not expect as a message with a different structure. We will go through how to mitigate all of the difficulties, and we will start by showing how to communicate with an external API using the native fetch method.

The fetch interface allows you to make HTTP requests to servers from our backend. It's a good choice for API communication for several reasons:

- **Native support**: fetch is natively available in modern browsers, eliminating the need for external libraries for basic request/response handling

- **Promise-based**: It works with Promises, enabling a more straightforward way to handle asynchronous operations and making the code cleaner and more readable

- **Support for various data types**: fetch can easily handle various types of data, including JSON, text, FormData, and even binary data such as ArrayBuffer and Blob, directly aligning with the needs of diverse APIs

Let's see fetch in action and how we can use it to define a function that is going to call the OpenAI API to generate GPT responses:

src/integrations/api.ts

```
export async function callGPTAPI(data: object) {
  const res = await fetch("https://api.openai.com/v1/chat/
completions", {
    method: "POST",
    body: JSON.stringify(data),
    headers: {
      "content-type": "application/json",
      authorization: "Bearer " + Bun.env.OPENAPI_API_KEY,
    },
  });
  return res;
}
```

This code defines an asynchronous function, `callGPTAPI`, which is intended to interact with an external OpenAI GPT API to obtain completions based on the data it sends. We will discuss more GPT functionalities in *Chapters 11* and *12*, but here, let's focus on the integration code:

- `fetch` returns a `Promise` object that resolves when you get the response to that request, whether it is successful or not.

- `method: "POST"` specifies the HTTP method the request uses; as we create something in this request, it's natural we use `POST`.

- `body: JSON.stringify(data)` is where the `data` object is converted into a JSON string because the API expects the request payload to be in JSON format. This is necessary because the `fetch` API does not automatically convert objects into JSON strings. The `data` object contains parameters for the API request, such as the prompt and other settings for the completion request.

- `"content-type": "application/json"` tells the server that the body of the request contains JSON. This is important for the server to correctly interpret the data sent in the request.

- `Authorization: "Bearer " + Bun.env.OPENAPI_API_KEY` provides the necessary authentication to access the API. The API uses a similar authentication method of a JWT token as our backend. We will talk about how to create an `OPENAPI_API_KEY` token in the next chapter.

This basic configuration is designed for handling API integration when it succeeds, commonly known as the "happy path." We also need to handle cases of failures from the APIs, so let's talk more about error management and retries.

Handling errors and retries

We should always treat APIs as erroneous because from time to time, they will fail, and we will need to be ready for that. One popular solution for solving intermittent external API errors is to try the request again with an increasing time between calls, a technique called **retries with exponential backoffs**. A retry with exponential backoff means if a request fails, we are going to try calling the external API again and wait more and more time before each call. Let's implement a function that is going to be a wrapper for our API call to enhance our API call with retries and exponential backoff:

src/integrations/retry.ts

```
export async function retryWrapper(
    fn: () => Promise<Response>,
    retryCount: number = 3,
    delayMs: number = 1000,
): Promise<Response> {
  async function attempt(attemptNumber: number = 1): Promise<Response>
```

```
{
    try {
      const result = await fn();
      if (!result.ok) {
        throw new Error(`Request failed with status ${result.
status}`);
      }
      return result;
    } catch (error: unknown) {
      if (retryCount > 0) {
        await new Promise((resolve) => setTimeout(resolve, delayMs *
Math.pow(2, attemptNumber - 1)));
        retryCount--;
        return attempt(attemptNumber + 1);
      } else {
        throw new Error(
            `API calls failed after retries: ${(error as Error)?.
message}`,
        );
      }
    }
  }
  return attempt();
}
```

Let's explore in detail what is happening in the function:

- Function parameters:

  - fn: The function to execute, which returns a Promise<Response> instance. This is the main action you're trying to perform – our API request.

  - retryCount: The number of times to retry fn upon failure before giving up. Defaults to 3.

  - delayMs: The initial delay in milliseconds before the first retry attempt, in case of failure. This delay increases exponentially in subsequent retries. Defaults to 1000 milliseconds (1 second).

- Inner attempt function:

  - An asynchronous function defined within retryWrapper that attempts to execute fn.

  - It takes attemptNumber as a parameter, starting from 1, to track the number of attempts made and to calculate the delay for the exponential backoff.

- Executing fn and handling success:

  - The fn function is called within a `try` block. If the call is successful (`result.ok` is `true`), a response is returned immediately.

  - If the response indicates failure (`result.ok` is `false`), an error is thrown, prompting a retry attempt.

- Handling failures and retries:

  - In case of an error (either from fn failing directly or the response being unsuccessful), the `catch` block checks if there are remaining retries with `retryCount > 0`.

  - Before retrying, the function waits for a period calculated by `delayMs * Math.pow(2, attemptNumber - 1)`. This wait time increases exponentially with each attempt, based on `attemptNumber`, therefore, the wait time becomes 1, 2, and 4 seconds. This is the exponential backoff mechanism, aimed at reducing the load on the server and improving the chance of success in subsequent attempts.

  - `retryCount` is decremented, and `attempt` is recursively called with `attemptNumber + 1` to indicate the next attempt.

- Exhaustion of retries:

  - If `retryCount` reaches 0, meaning all retry attempts have been exhausted without success, the function throws an error, indicating that the API calls failed after the specified number of retries.

  - The original error message from the last attempt is included in the error message for debugging purposes.

- Return value:

  - The `retryWrapper` function initiates the retry logic by calling `attempt` and returns its result. This will either be a successful response from fn or an error after all retries have been exhausted.

This retry mechanics is going to ensure that we are going to increase our chance of solving short temporary issues with the external API, and it will throw an error if it eventually doesn't happen. Now, we can move on to validating the response we got.

Validating API correctness

We have no guarantees on what the API is going to return to us. Or, rather, we do, but it's a bad practice to rely solely on guarantees, especially if the API is not controlled by us. To ensure that what we expect is what we need to validate the response from the API call, our code will use zod to ensure that responses from an API match a predefined structure, effectively validating the integrity of the data before it's used in the application. If the data is correctly structured, the function extracts and returns a specific piece of content; otherwise, it throws an error indicating what went wrong:

src/integrations/validation.ts

```
import { z } from "zod";

const GPTResponseSchema = z.object({
  choices: z.array(
    z.object({
      message: z.object({
        content: z.string(),
      }),
    }),
  ),
});

export async function validateGPTResponse(response: Response):
Promise<string> {
  const responseData = await response.json();
  try {
    const parsed = GPTResponseSchema.parse(responseData);
    const content = parsed.choices[0].message.content.trim();
    return content;
  } catch (error) {
    throw new Error(`Invalid API response format, format ${error}`);
  }
}
```

Let's break down the components of the function :

- First, we import the zod library.

- Then, we define GPTResponseSchema. According to this schema, the response should be an object with a key named choices, which is an array of objects. Each object within the choices array should have a message object, which in turn should have a content string.

- This structure mirrors a typical structure you might expect from a GPT API response, where `choices` contains an array of possible completions or responses generated by the model.

- The `validateGPTResponse` function is an asynchronous function that takes a `Response` object (the result of a `fetch` call) as its argument.

- It first converts `Response` to an object using `response.json()`.

- It then attempts to validate this object against `GPTResponseSchema` defined earlier using `.parse(responseData)`. If the data matches the schema, it means the API response has the expected format and the function proceeds.

- The code extracts the content of the first choice's message, trims any leading or trailing whitespace using `.trim()`, and returns this trimmed content as a string.

- If the response data does not match `GPTResponseSchema`, zod will throw an error during the `.parse()` call.

- This error is caught in the `catch` block, and the function then throws a new error with a message indicating that the API response format is invalid. The original error message from zod (which typically contains details about what part of the data did not match the schema) is included in the new error message for debugging purposes.

Now, we can retry our API call and also validate the response, so let's put all the pieces together in our function invocation:

src/integrations/gpt.ts

```
import { HTTPException } from "hono/http-exception";
import { callGPTAPI } from "./api";
import { retryWrapper } from "./retry";
import { validateGPTResponse } from "./validation";

export async function getGPTAnswer(data: object) {
  try {
    const response = await retryWrapper(() => callGPTAPI(data));
    const message = await validateGPTResponse(response);
    return message;
  } catch {
    throw new HTTPException(503, { message: "GPT integration is down"
});
  }
}
```

Let's break this code down here:

- First, we include `HTTPException` from `hono/http-exception`, which we can use to return responses with custom error codes and messages we want. Then, we import all the functions we defined in the previous code pieces.

- `GetGPTAnswer` accepts the data that we are going to send further to `callGPTAPI` and returns the GPT-generated message or an `HTTPException` error with code `503`, which means that our server is temporarily unavailable.

- Finally, we pass an anonymous function to `retryWrapper`, which is going to call our API with `() => callGPTAPI(data)`, and then we try to validate the response and return the result.

With this in place, we are ready to securely and reliably call our GPT API. Our API is not callable yet, and we will explore in the next two chapters how to finish it.

Summary

In this chapter, we tackled the crucial task of showing how to integrate an API with our backend server, spotlighting the utility of `fetch` for executing HTTP requests within the Bun environment. We navigated through enhancing API reliability by addressing error handling, implementing retries with exponential backoff, and ensuring the integrity of API responses through structured validation. These strategies collectively improve our API interactions, enabling the development of secure, robust, and efficient backend integrations.

The next chapter will delve into setting up and configuring the OpenAI API so that we can call it from our application. This will include generating API keys to be able to call the OpenAI API and then finishing the integration to generate a GPT response end to end.

Setting Up and Configuring the OpenAI API for the Backend

In this chapter, we'll explore and master the integration of the OpenAI API into our backend setup. Starting with a deep dive into **large language models** (**LLMs**) and the OpenAI API, we'll cover everything from the basics to practical implementation steps. This knowledge is not only important for enhancing the technical repertoire with the cutting-edge capabilities of **artificial intelligence** (**AI**) but also for leveraging these technologies to innovate and streamline processes in your applications. By the end of this chapter, you'll be well equipped with the skills to configure secure and efficient OpenAI API integration and bring the power of AI into your TypeScript projects.

In this chapter, we're going to discuss the following topics:

- Introduction to LLMs and their applications
- Setting up OpenAI API integration
- Integrating the OpenAI API into our backend

We will start with a general introduction to LLMs and their applications.

Technical requirements

For this chapter, we won't need any additional libraries.

All the code examples we discuss are available in the GitHub repository: `https://github.com/ PacktPublishing/Full-Stack-Web-Development-with-TypeScript-5/tree/ main/Chapter11`

Introduction to LLMs and their applications

LLMs are a big step forward in AI, helping machines understand and create text that sounds like it was written by a human, thanks to a lot of training data. At the core of these developments are **generative pre-trained transformers** (**GPTs**), a group of models recognized for their outstanding capability to produce text that is both coherent and contextually appropriate on a variety of topics.

LLMs are at the forefront of AI, enabling computers to understand and produce text that mirrors human communication. These models are trained on extensive datasets, covering a broad spectrum of human knowledge, to predict and generate text based on input. Among the most prominent LLMs is OpenAI's GPT, which has revolutionized the way machines understand and interact with human language. Let's break down the technical workings in a more digestible manner:

- **Transformer architecture**: GPT models leverage the transformer architecture for parallel word processing, moving away from traditional sequential processing. They feature a self-attention mechanism that lets the model assess the importance of all words in a sentence at once, improving its understanding of context.

- **Pre-training and fine-tuning**: Initially, GPT models are pre-trained on vast datasets to grasp general language patterns. This stage is unsupervised, focusing on next-word prediction. They're then fine-tuned with specific datasets to tailor the model for particular tasks such as translation or question-answering.

- **Tokenization and embeddings**: The input text is segmented into tokens (words or subwords), which are then transformed into numerical embeddings. These embeddings encode semantic information, facilitating language generation and understanding.

- **Attention mechanism**: Utilizing the self-attention mechanism, the model assesses the entire input to decide on the significance of each word when generating text, allowing it to draw connections across different parts of the text.

- **Layered structure**: Comprising multiple layers of transformer blocks, GPT models progressively refine text representation through each layer, enabling the handling of complex language structures.

- **Output generation**: The model generates text by predicting the most likely next word based on the input and its learned patterns, continuing until a specified end condition is met.

Integrating GPT models via the OpenAI API also involves understanding several key technical aspects to effectively use its capabilities. Here's a breakdown of the main concepts:

- **Tokens**: Tokens are the basic units of text that the model processes. Input text and output generation are measured in tokens, which can be words, parts of words, or punctuation. The total number of tokens you can send in a single request is limited, so understanding how to effectively use tokens is important for optimizing API calls and managing longer conversations or documents. One token is approximately four characters.

- **Context windows**: The context window refers to the maximum number of tokens the model can consider for generating a response. Each version of GPT limits the context window size, impacting how much previous text the model can consider when generating its output.

- **Hallucinations**: Hallucinations refer to instances where the model generates factually incorrect or nonsensical information. While GPT models are highly advanced, they can still produce errors or fabricate details, especially when dealing with topics outside their training data or when pushed beyond their limits of understanding. Handling hallucinations often involves refining prompts or postprocessing the output to ensure accuracy. At the same time, it's important to understand that hallucinations are essentially what LLMs do all the time, as they simply try to guess the next word that "makes sense" in the given context.

- **Temperature**: Temperature controls the randomness of the model's output. A lower temperature results in more deterministic and predictable text, while a higher temperature encourages creativity and variability. Modifying the temperature parameter allows you to customize the model's responses based on your requirements, from generating highly creative content to providing accurate information.

- **Prompt engineering**: Crafting effective prompts is an art on its own, significantly influencing the model's output. Good prompt engineering involves providing clear, concise instructions and context to guide the model toward the desired output. It can include specifying the format, style, or content of the response and is a key skill for maximizing the potential of GPT integration.

Knowing this, we can discuss the technical cases that LLMs and GPTs can help us to solve, as the possibilities are vast:

- **Content creation**: Automating the generation of articles, reports, and narratives, saving time and resources

- **Chatbots**: Developing sophisticated virtual assistants that provide human-like interactions for customer support and engagement

- **Code generation**: Assisting developers by generating code snippets, thus speeding up the development process

- **Language translation**: Facilitating real-time, accurate language translation services for global applications

- **Personalized recommendations**: Generating personalized content or product recommendations in e-commerce platforms

- **Data analysis**: Summarizing complex datasets into understandable reports, enhancing decision-making processes

- **Educational tools**: Creating personalized learning materials and interactive learning environments

- **Email automation**: Drafting and personalizing email responses, improving communication efficiency

- **Search engines**: Enhancing search algorithms to understand **natural language** (**NL**) queries, delivering more relevant results

- **Game development**: Generating dynamic dialogues and narratives, creating more immersive gaming experiences

- **Classification**: GPT models can decide on a category of information and be used to, for example, detect spam

There are also many other players in the field that provide APIs to their models. One of the prominent examples at the time of writing this book is Claude 3 Opus, which allows a much bigger context window than GPT-4. This means that Claude 3 Opus allows the acceptance of more data as input and produces more data as output, which is required in certain cases. In the next section, we will showcase an example on the OpenAI API, but the integrations and concepts are mostly similar to other providers.

Let's now see how we can configure OpenAI to get an API token for calling models' endpoints from our code.

Setting up OpenAI API integration

To use the GPTs provided by OpenAI, we need to set up our account there and create an API token that will make it possible for us to access models programmatically. To do so, follow these steps:

1. First, we will need to register here: `https://platform.openai.com/signup`. Now, we need to add a payment method and top up our account with $5 for the experiments.

2. We need to add a payment method here by clicking on the **Add payment details** button: `https://platform.openai.com/settings/organization/billing/overview`.

3. You then need to enter payment information. After this, you will be prompted to add credits to the account.

4. Type the minimum amount allowed; it will be enough for our experiments. Now, we can start creating an API key.

5. To do this, we will go to the **Api keys** section: `https://platform.openai.com/api-keys`.

6. We will be asked to also validate our phone number by a notification at the top of the screen.

7. Now, we need to create a new secret key by clicking on **Create new secret key**.

8. In the popup, we will go to the **Restricted Permission** tab.

9. We will enable access to chat completion by enabling only the section with **/v1/chat/completions** by setting **Write** to it.

10. You will get a secret key generated that you will never see again, so keep it secure.

We've got the API key that we will use in our application. Now, let's finish up our integration with OpenAI from our code and see how it works.

Integrating the OpenAI API into our backend

We wrote most of the code for our API integration in the previous chapter, in the *Introduction to API integration in TypeScript using fetch* section, so now it is time to write the missing pieces.

First, let's put the API key that we got into our environment file:

.env

```
OPENAPI_API_KEY=HERE_GOES_YOUR_KEY
```

Now, we can add a function that will call our API integration code with configurations for a GPT model.

Here, we import the `DBMessage` type for our message and the `getGPTAnswer` function we created in the previous chapter to call GPT:

src/integrations/generate_message.ts

```
import type { DBMessage } from "../models/db";
import { getGPTAnswer } from "./gpt";
```

Now, we can define a `generateMessageResponse` function that is going to generate a GPT message response to the user message:

```
export async function generateMessageResponse(
  messages: DBMessage[],
): Promise<string> {
```

Here, we accept all messages relevant to our conversation to generate the next one. It is important to include the message history and not only the last message, so that we have a consistent conversation. Let's configure the parameters needed for our GPT model:

```
const params = {
  model: "gpt-3.5-turbo-0125",
  temperature: 0,
  max_tokens: 1000,
  top_p: 1,
  n: 1,
  stream: false,
  stop: "",
};
```

Let's discuss the parameters:

- `model`: Specifies the version of the GPT model to use for the request. `"gpt-3.5-turbo-0125"` indicates a specific iteration of the GPT-3.5 Turbo model, optimized for certain types of tasks and performance characteristics.

- `temperature`: Controls the randomness of the output. A temperature of 0 makes the model's responses deterministic, meaning given the same prompt, the model will always generate the same response. The max value is 1, which makes it as "creative" as possible. This setting is useful for tasks requiring consistency and less creativity.

- `max_tokens`: Sets the maximum number of tokens (including both the prompt and the generated response) that the model will output in its response. 1000 tokens means the response, combined with the input prompt, will not exceed 1,000 tokens in total. This limit is important for controlling the length of the generated content.

- `top_p`: Known as "nucleus sampling," `top_p` controls the number of potential words the model evaluates. A higher `top_p` value allows the model to consider a wider range of words, including those that are less probable, resulting in more varied text output. Setting it to the max value of 1 means all potential words are included in the generation, and setting it to 0.5 means only words that make up 50% of all the words' probability will be used.

- `n`: Specifies the number of responses to generate for the given prompt. 1 means the model will generate a single response for each request. This is useful for applications that require a unique output for each input.

- `stream`: Determines whether the model's output should be streamed. `false` indicates that a response will be sent back only after it's fully generated. Streaming is beneficial for receiving the model's output in real time, especially for longer responses, but here it is not enabled.

- `stop`: Sets a stop sequence at which the model will stop generating further tokens. An empty string (`""`) means there is no specific stop sequence, and the model will continue generating tokens until another stopping condition is met (such as reaching the `max_tokens` limit).

Now, let's construct the `data` object we are going to send to our OpenAI endpoint:

```
const data = {
  ...params,
  messages: [
    {
      role: "system",
      content: `You are a helpful AI assistant who answers to the
user messages`,
    },
    ...messages.map((m) => ({ role: m.type, content: m.message })),
  ],
};
```

```
    return getGPTAnswer(data);
}
```

Here, we create a `data` object that consists of configuration parameters and messages we are going to send to the API. The first message by the "`system`" role is the way we provide additional instructions for our GPT models, as it listens to the "`system`" role as an instruction.

At the end, we return a response from calling `getGPTAnswer`.

We are almost done; the only thing left is to actually call the `generateMessageResponse` function from our message creation endpoint. For this, we will need to replace our dummy message with a call to this function:

src/controllers/chat.ts

```
import { generateMessageResponse } from "../integrations/generate_
message";
...
chatApp.post(
    CHAT_MESSAGE_ROUTE,
...

    await messageResource.create(userMessage);

    const allMessage = await messageResource.findAll({ chatId });
    const response = await generateMessageResponse(allMessage);
    const responseMessage: DBCreateMessage = {
      message: response,
      chatId,
      type: "user",
    };
```

First, we get all the messages in the chat, then we get the response from our endpoint, and finally, we put it in our DB message object to insert in the database and return to the user.

The only bit left is that now, as we call the real endpoint, we need to also adapt our test to account for it. Let's change the lines where we expect the response message to be a dummy message to any non-empty string:

tests/chat.test.ts

```
...
expect(messages.data[0].message).toBe("Hello World");
expect(messages.data[1].message?.length).toBeGreaterThan(0);
```

With this in place, our integration is ready, and now, if you call the message creation endpoint from `curl`, you will observe that we get a GPT-generated response in return.

This has been a great feat, and now we are done with the backend part of the functionality and are ready to get on the frontend journey.

Summary

In this chapter, we covered how LLMs and GPT models work and the main technical aspects we need to know to integrate with them. Then, we discussed how to set up an account at OpenAI and get a token for API calls. Finally, we integrated the API into our backend system. All this knowledge enables us to integrate AI-based functionality so that we can solve various tasks in our backend systems.

Now, we are going to turn our attention to the frontend part of our chat application and our Svelte journey.

Part 5:
Frontend Development
with Svelte

The final part of the book focuses on frontend development using **Svelte**, a modern tool for building reactive and efficient user interfaces. This part will guide you through setting up a Svelte project, developing a chat application, and applying advanced frontend techniques to enhance performance and maintainability. It's designed to provide a comprehensive understanding of how to build and optimize frontend architectures effectively.

This part includes the following chapters:

- *Chapter 12, Introduction to Svelte for Frontend Development*
- *Chapter 13, Setting up the Svelte Project*
- *Chapter 14, Svelte Chat Application Development*
- *Chapter 15, Advanced Svelte Techniques*

12

Introduction to Svelte for Frontend Development

We've successfully finished our backend logic and learned how to build robust and extensible APIs. Now, it's time for us to write the part of the application that matters the most for our end users: the frontend. To do that, we will use Svelte as our frontend library. Before we start building the actual application, we will need to discuss what Svelte is, why we chose it, and cover the main aspects of developing apps in Svelte – and this is exactly what we are going to cover in this chapter. By the end of it, we will have a good grasp of Svelte and will be equipped with the knowledge to start developing the frontend with it.

We will cover the following topics in this chapter:

- What is Svelte?
- Learning Svelte fundamentals

Technical requirements

To proceed with this chapter, we won't need any additional libraries.

All the code examples we are going to discuss in this chapter are available at `https://github.com/PacktPublishing/Full-Stack-Web-Development-with-TypeScript-5/tree/main/Chapter12`.

What is Svelte?

Svelte is an innovative component-based JavaScript framework for building UIs. Unlike traditional frameworks such as React, Angular, or Vue, Svelte shifts much of the work from the browser to the build step, compiling applications to highly optimized vanilla JavaScript at build time.

Let's touch briefly on its history now and also understand how it is different from other frameworks.

History of Svelte

Svelte was created by Rich Harris and first released in 2016. Harris, a graphics editor at *The New York Times*, designed Svelte to address the complexities and performance bottlenecks inherent in existing JavaScript frameworks. Svelte has evolved rapidly, with Svelte 3 introduced in 2019, bringing significant advancements, including a new reactivity model that further distinguishes it from its predecessors. Svelte 4 was released in 2023 with major performance improvements.

Differentiation from other frameworks

The key differentiator of Svelte lies in its compile-time philosophy. While frameworks such as React and Vue rely on virtual DOM diffing algorithms at runtime to update the UI, Svelte generates minimal, imperative code during build time to directly update the DOM when the state of an app changes. This approach reduces overhead and runtime dependencies, leading to faster initial loads and smoother updates.

Pros of using Svelte include the following:

- **Performance**: By avoiding the virtual DOM and minimizing runtime overhead, Svelte applications typically start faster and remain responsive even as they grow in complexity.

- **Simplicity**: Svelte's syntax is clean and approachable with less boilerplate, making it easier for developers to read and write code. Its reactivity model is straightforward, relying on assignment to trigger updates.

- **Developer experience**: Svelte offers a delightful developer experience, with clear error messages, comprehensive documentation, and a supportive community. Its single-file component format integrates styles, logic, and markup, streamlining development.

- **Size efficiency**: Applications built with Svelte are often smaller than those built with other frameworks, thanks to the compile-time optimizations. This results in better performance and faster loading times.

- **Built-in features**: Svelte comes with built-in support for animations, transitions, and state management, reducing the need for external libraries.

- **Ease of learning**: Svelte introduces a linear learning curve and provides a tremendously useful interactive tutorial into all of its features at `https://learn.svelte.dev/tutorial`.

While Svelte offers many advantages, there are also some considerations that might make it less suitable for certain projects compared to other established frameworks. Here are some of the cons associated with Svelte:

- **Community and ecosystem size**: Being newer and less widespread than giants such as React or Angular, Svelte's community and ecosystem are smaller. This can mean fewer third-party libraries, resources, and community support available, which might hinder solving specific problems or finding pre-built solutions.

- **Fewer learning resources**: Although the quality of Svelte's official documentation is excellent, there might be fewer tutorials, courses, and external learning materials available compared to more established frameworks.

- **Job market**: The demand for Svelte developers is growing but still lags behind the demand for developers with expertise in more established technologies such as React, Vue, or Angular. This could be a consideration for developers prioritizing the marketability of their skills.

- **Advanced use cases**: While Svelte is highly capable, very complex applications might benefit from the more mature ecosystems of larger frameworks, which offer a wider range of solutions for scalability and optimization out of the box.

- **Tooling and integrated development environment (IDE) support**: Though improving, tooling around Svelte (for example, plugins for code editors, IDEs, and debugging tools) may not be as advanced or as plentiful as those available for more established frameworks.

Despite these considerations, there are compelling reasons why Svelte was chosen for developing our frontend application. Svelte is driven by its performance benefits, the developer experience it offers, and its suitability for the project's specific requirements. Despite its smaller ecosystem, Svelte's advantages align well with our goals of building a fast, efficient, and maintainable frontend application.

It's time that we get to know Svelte with examples.

Learning Svelte fundamentals

We will start our journey by talking about what we can see in a Svelte component and how it works together.

Component composition structure

Svelte components are encapsulated units of HTML, CSS, and JavaScript (or TypeScript) logic that define parts of a UI. A **Svelte component** is typically written in a `.svelte` file, which neatly bundles together the structure (HTML), appearance (CSS), and behavior (JavaScript/TypeScript) of UI elements. When using TypeScript, you can define types and interfaces to ensure type safety and improve the development experience.

Introduction to .svelte file structure

A `.svelte` file is divided into three primary sections:

1. **HTML/markup block**: This section contains the HTML structure of the component and Svelte-specific syntax for reactivity, loops, conditionals, and event handling, enclosed within the `<html>` tag.

2. **Script block**: Enclosed within `<script>` tags, this area is where you write the component's JavaScript or TypeScript logic. To use TypeScript, we will need to include `lang="ts"` in the `<script>` tag. In this tag, you can define component props, local state, functions, and reactive statements.

3. **Style block**: CSS styles are placed within a `<style>` tag. Styles defined here are scoped to the component, meaning they won't leak out and affect other parts of the application unless explicitly made global.

Let's see an example of a simple component with all these mentioned parts:

Hello.svelte

```
<script lang="ts">
  export let name: string = 'World';
</script>

<main>
  <h1>Hello {name}!</h1>
</main>

<style>
  main {
    text-align: center;
    color: purple;
  }
</style>
```

In Svelte, it is more common to stick to the Pascal case names, so this is what we used for the filename here:

1. **Script block**: The component begins with a script block where TypeScript is enabled, `<script lang="ts">`. We declare a prop named `name` of type `string`, which allows this component to receive a name value from its parent.

2. **HTML/markup block**: The HTML structure consists of a `<main>` element containing an `<h1>` tag. The text within the `<h1>` tag dynamically displays the `name` prop's value, showcasing Svelte's reactive bindings. Curly braces (`{name}`) are used to reference the JavaScript variable in the HTML markup.

3. **Style block**: The styles are scoped to this component, styling the `<main>` element to center the text and set its color to purple. These styles won't affect elements outside this component.

Let's now cover how reactivity works in Svelte.

Reactivity

Svelte's reactivity system is uniquely straightforward and powerful, enabling seamless updates to the UI in response to state changes with minimal code. By simply assigning new values to variables, Svelte components reactively update the DOM. Additionally, Svelte's reactive statements, denoted by $:, allow for automatic re-computation of expressions when their dependencies change.

Let's explore both concepts—reactivity through assignment and reactive statements—within a Svelte example. We will create a counter that we can increment when you click on a button:

Count.svelte

```ts
<script lang="ts">
    let count: number = 0;
    $: doubled = count * 2;
    function increment() {
        count += 1; // Triggers reactivity
    }
</script>

<button on:click={increment}>
    Clicked {count} {count === 1 ? 'time' : 'times'}
</button>
<p>Doubled Value: {doubled}</p>
```

Let's discuss the preceding code block in detail:

- Script block:

  - The count variable is declared with a type of number and is initialized to 0. This simple state will be reactively updated in the UI whenever its value changes.

  - A $: doubled = count * 2; reactive statement is used to define a derived value named doubled, which automatically recalculates whenever count changes. This showcases how effortlessly Svelte handles derived state or side effects in response to state modifications.

  - The increment function demonstrates triggering reactivity in Svelte. By incrementing count, we not only update the count variable itself but also indirectly cause doubled to update, thanks to the reactive statement.

- HTML/markup block:

  - Contains a button that displays the current count. Each click on the button calls `increment`, showcasing the direct reactivity by updating the text displayed on the button itself.

  - The `<p>Doubled Value: {doubled}</p>` paragraph displays the derived `doubled` value, which updates reactively as `count` changes, illustrating the power of reactive statements for managing derived state.

We've seen `click` events in action in this example of code, but let's explore further how to write and handle events in Svelte.

Events

Events in Svelte are handled through the `on:eventName` directive, making it straightforward to listen for user actions such as clicks, input changes, or more complex events. Modifiers such as `preventDefault` can be appended to event directives to modify the event's default behavior directly in the markup, streamlining the code.

Components can dispatch their own custom events using Svelte's built-in `createEventDispatcher` function. This feature is particularly useful for creating encapsulated, reusable components that communicate with their parents or other components.

In this class, we will increment a counter when a user clicks on the button. We will reset the state to 0 when the **Reset** button is pressed, and we are also going to dispatch a custom event to the parent:

EventDispatch.svelte

```ts
<script lang="ts">
    import { createEventDispatcher } from 'svelte';

    let count: number = 0;
    const dispatch = createEventDispatcher<{ reset: void }>();

    function increment() {
        count += 1;
    }

    function handleResetClick(event: MouseEvent) {
        count = 0;
        dispatch('reset'); // Dispatching a custom event named 'reset'
    }
</script>
```

The counter's state is stored in a `count` variable. The `increment` function increases this `count` variable, demonstrating direct event handling by incrementing the `count` variable upon a button click.

We use Svelte's `createEventDispatcher` function to create a dispatcher function, enabling the component to emit a custom `reset` event. The `handleResetClick` function resets the `count` variable to `0` and dispatches this event, showcasing how components can communicate actions upward to their parent components or other listeners.

The `handleResetClick` function also demonstrates how TypeScript can be used to type event handlers, in this case, specifying that the `event` parameter is of type `MouseEvent`.

Two buttons are defined: one for incrementing the counter and another for resetting it. An `on:click` directive is used to attach event handlers to these buttons:

```
<button on:click={increment}>Increment</button>
<button on:click|preventDefault={handleResetClick}>Reset</button>
```

The `preventDefault` modifier is used with the **Reset** button's `click` event to demonstrate event modifiers in Svelte. This modifier prevents the default action that belongs to the event (useful in cases such as form submission buttons within `<form>` tags), although its effect is not directly observable in this example since the button click doesn't have a default behavior to prevent.

We can now write a parent component that is going to handle the custom `reset` event we wrote:

EventCatch.svelte

```
<script lang="ts">
    import EventDispatch from './EventDispatch.svelte';
    function handleReset() {
        console.log('Counter was reset');
    }
</script>

<main>
    <EventDispatch on:reset={handleReset} />
</main>
```

The `handleReset` function is defined to perform actions when the `reset` event is caught. This is where you can add any logic that should execute in response to the `reset` event from the `EventDispatch` component.

In the markup, we include the `Counter` component and use the `on:reset` directive to listen for the `reset` event. When the `reset` event is dispatched from the `Counter` component, the `handleReset` function is called.

Next, let's cover another important aspect of how to handle form data with the help of bindings in Svelte.

Bindings

Svelte simplifies the synchronization of UI elements with application state using its intuitive binding system. This system allows for a two-way data flow between state variables and UI elements, such as input fields and select boxes, without the need for explicit event handling or manual DOM updates. Let's explore this with an example that demonstrates bindings with an input field and a select box.

In this example, we'll create a Svelte component that includes a text input and a select box. Both will be bound to local state variables, showcasing how changes to these UI elements automatically update the variables and vice versa.

Two state variables, name and favoriteColor, are declared with initial values. These variables will be bound to an input field and a select box, respectively:

Binding.svelte

```ts
<script lang="ts">
    let name: string = 'John Doe';
    let favoriteColor: string = 'blue';

    const colors: string[] = ['red', 'green', 'blue', 'yellow'];
</script>
```

A colors array is defined, listing the options for the select box. This array is used to dynamically generate the select box options using Svelte's {#each} block.

The text input is bound to the name variable using bind:value={name}:

```
<div>
    <input type="text" bind:value={name}/>
```

This creates a two-way binding between the input's value and the name variable, meaning any change in the input field will automatically update name, and any programmatic change to name will update the input field's value.

The select box is bound to the favoriteColor variable using bind:value={favoriteColor}:

```
<select bind:value={favoriteColor}>
```

Similar to the input binding, this establishes a two-way binding, ensuring the select box reflects the current favoriteColor value and updates it upon user selection.

The select box options are dynamically generated using Svelte's {#each} loop, iterating over the colors array:

```
{#each colors as color}
    <option value={color}>{color}</option>
```

```
    {/each}
  </select>
```

This approach demonstrates how to handle dynamic lists in select boxes with Svelte.

This demonstrates the reactivity of Svelte bindings by displaying the current values of `name` and `favoriteColor`. As these variables are updated via the UI elements they're bound to, the displayed text automatically updates to reflect the current state:

```
</div>
```

```
<p>Hello, {name}! Your favorite color is {favoriteColor}.</p>
```

We can continue with conditionals and look more into how to handle arrays in Svelte.

Handling conditionals and iterating arrays

Svelte provides a straightforward syntax for conditional rendering and iterating over lists, making dynamic UIs simpler to implement. Let's explore how to use Svelte's `{#if}`, `{#else}`, and `{#each}` blocks, combined with TypeScript, to control what is rendered based on the application's state and to display lists of items.

Our example will demonstrate creating a task list where tasks can be marked as completed. It will showcase conditional rendering to display different messages based on whether the list is empty and iterating over a list of tasks with the ability to dynamically update their completion status:

Conditional.svelte

```
<script lang="ts">
    interface Task {
        id: number;
        name: string;
        completed: boolean;
    }
```

A `Task` interface is defined to describe the structure of a task, ensuring type safety for our list of tasks:

```
    let tasks: Task[] = [
        {id: 1, name: 'Learn Svelte', completed: false},
        {id: 2, name: 'Build an app', completed: false},
    ];
```

The `toggleTaskCompletion` function toggles the completion status of a task. It uses `taskId` to find the task in the array and updates its `completed` property:

```
function toggleTaskCompletion(taskId: number) {
```

The `if` statement checks if the `tasks` array is empty. If so, a message prompts the user to add tasks:

```
    const taskIndex = tasks.findIndex(task => task.id === taskId);
    if (taskIndex !== -1) {
        tasks[taskIndex].completed = !tasks[taskIndex].completed;
    }
}
</script>
{#if tasks.length === 0}
<p>No tasks found. Add some tasks!</p>
```

This illustrates how to conditionally render content based on the application's state.

Now, we will see an example of using Svelte's `else` block to handle alternative content rendering.

In the next lines, we will see a `{#each}` block that iterates over the `tasks` array, rendering a `<li>` list item for each task. This shows how to display dynamic lists in Svelte:

```
{:else}
    <ul>
        {#each tasks as task}
            <li on:click={() => toggleTaskCompletion(task.id)}>
                <input type="checkbox" bind:checked={task.completed}/>
{task.name}
            </li>
        {/each}
    </ul>
{/if}
```

Each list item contains a checkbox bound to the task's `completed` property, demonstrating two-way data binding within a list. Clicking the list item calls `toggleTaskCompletion` for the task, toggling its completion status. This interaction showcases handling events and updating state based on user actions.

We can now talk about how props passing works in Svelte.

Props

Props in Svelte are a way to pass data from a parent component to a child component, enabling component reuse and the composition of complex interfaces. When using TypeScript, you can enhance props with types, ensuring that components receive data in the expected format and improving the developer experience through better tooling support.

To declare a prop in a Svelte component, you use the `export` keyword on a variable within the `<script>` tag. This makes the variable accessible to any parent component that uses the child component. TypeScript allows you to specify the type of each prop, adding a layer of type safety.

Our example demonstrates creating a `Greeting` component that accepts a `name` prop of type `string`. The parent component will pass a name to the `Greeting` component, which will then render a personalized message:

Greeting.svelte

```ts
<script lang="ts">
  export let name: string;
</script>

<p>Hello, {name}!</p>
```

The name variable is declared and exported, making it a prop that the `Greeting` component expects from any parent component. Specifying the type as `string` ensures that `name` is treated as a string, leveraging TypeScript's type checking.

The markup simply displays a greeting message incorporating the `name` prop.

Now, let's use the `Greeting` component within a parent component, passing a name to it as a prop:

GreetingParent.svelte

```ts
<script lang="ts">
    import Greeting from './Greeting.svelte';
</script>

<main>
    <Greeting name="John" />
</main>
```

Within the parent's `<main>` element, the `Greeting` component is instantiated and given a `name` prop with the value `"John"`. This showcases how data is passed down from parent to child components in Svelte.

Now, let's cover the lifecycle of a Svelte component.

Lifecycle

Svelte offers lifecycle hooks that allow you to perform actions at different stages of a component's life, such as when it's created, updated, or destroyed. These hooks are crucial for managing resources, subscriptions, and other side effects in a component.

Let's create a component that utilizes lifecycle hooks to demonstrate how they can be used:

Lifecycle.svelte

```ts
<script lang="ts">
    import { onMount, beforeUpdate, afterUpdate, onDestroy } from
'svelte';
    let count: number = 0;
    onMount(() => {
        console.log('Component mounted');
    });
    beforeUpdate(() => {
        console.log('Component will update');
    });
    afterUpdate(() => {
        console.log('Component updated');
    });
    onDestroy(() => {
        console.log('Component will be destroyed');
    });

    function increment() {
        count += 1;
    }
</script>

<button on:click={increment}>Increment</button>
<p>Count: {count}</p>
```

Let's talk about the lifecycle hooks that we use here:

- onMount: This hook is used to log a message when the component is mounted. It's an ideal place for initialization tasks, such as fetching data or setting up subscriptions that require the component to be present in the DOM.

- beforeUpdate and afterUpdate: These hooks provide insight into the component's reactive updates. beforeUpdate is useful for preparations before the DOM updates, and afterUpdate can be used for actions that require the updated DOM, such as scrolling or focusing elements

- onDestroy: Here, a cleanup action is logged, demonstrating where you would typically free resources or remove event listeners to prevent memory leaks.

We are now ready to cover the last part, which is to handle the state of our app using stores.

Stores

Svelte stores are a built-in mechanism for managing state reactively across components. They're particularly useful for sharing stateful logic and data in different parts of your application without the need for prop drilling or context providers.

Svelte provides several types of stores, and the most commonly used is a `writable` store. A `writable` store allows you to create a reactive state that components can subscribe to and update from anywhere in the app.

Let's create a simple task manager where tasks can be added and marked as completed. We'll use a `writable` store to manage the tasks and TypeScript to ensure our tasks conform to a defined structure:

stores/taskStore.ts

```ts
import { writable } from 'svelte/store';

interface Task {
  id: number;
  title: string;
  completed: boolean;
}

const initialTasks: Task[] = [
  { id: 1, title: 'Learn Svelte', completed: false },
];

export const tasks = writable<Task[]>(initialTasks);
```

We define a `Task` interface that represents the structure of a task in our application.

A `writable` store named `tasks` is created with an initial array of tasks. The store is typed as `Task[]`, indicating it holds an array of `Task` objects.

Now, let's use the `tasks` store in a Svelte component to display and update a list of tasks:

Task.svelte

```ts
<script lang="ts">
  import { tasks } from './stores/taskStore';

  function toggleCompletion(taskId: number) {
    tasks.update(allTasks =>
      allTasks.map(task =>
        task.id === taskId ? { ...task, completed: !task.completed } :
```

```
task
        )
      );
    }
</script>

<ul>
   {#each $tasks as task}
     <li on:click={() => toggleCompletion(task.id)}>
        <input type="checkbox" bind:checked={task.completed} /> {task.
title}
     </li>
   {/each}
</ul>
```

We import the tasks store from its module. In Svelte, you access a store's value in markup by prefixing the store name with $.

The toggleCompletion function demonstrates how to update a store. It uses the update method provided by writable stores to toggle the completion status of a task. The update method takes a function that receives the current state and returns the updated state.

The {#each} block iterates over $tasks, rendering a list item for each task. The task's completion status can be toggled by clicking on the item, demonstrating a reactive update across components using the store.

With this, we have covered the basics of Svelte, and now we can proceed with coding our application in the next chapter.

Summary

In this chapter, we learned the fundamentals of Svelte, including file structure, reactivity, events, bindings, conditionals, arrays, props, and stores. All of this is going to be useful in the next chapter, where we are going to utilize the learned aspects to build a full-fledged chat application that integrates with our backend.

13

Setting Up the Svelte Project

Now that we have covered the basics of Svelte in *Chapter 12*, it is time for us to start developing our frontend application. We are going to cover how to set up our frontend **single-page application** (**SPA**) projects with Svelte, Vite, and TypeScript.

By the end of this chapter, we will have a solid grasp of how to initiate a project and how all the files of our frontend project play together.

We are going to cover the following topics in this chapter:

- Discussing Vite and SvelteKit
- Setting up the project
- Exploring the project structure

Technical requirements

To proceed with this chapter, we won't need any additional libraries.

All the code examples we are going to discuss in this chapter are available at `https://github.com/PacktPublishing/Full-Stack-Web-Development-with-TypeScript-5/tree/main/Chapter13`.

Discussing Vite and SvelteKit

Vite is a server for local development that we will use to serve our frontend application locally. Additionally, it is a build tool that we will use to ship our frontend application to production. Vite leverages the latest web technologies to offer instantaneous **hot module replacement** (**HMR**), which is a practice that only recompiles the parts of the application that we have changed) and extremely fast server startup times. This efficiency drastically reduces the feedback loop for developers, allowing for immediate observation of changes directly in the browser. Vite also has a plugin-based infrastructure that helps us to add extra functionality for it, such as integration with Svelte.

Now, let's talk about a popular framework for building Svelte apps that we chose not to go with in our app.

SvelteKit is a framework designed to build web applications using Svelte, offering features such as **server-side rendering (SSR)**, **static site generation (SSG)**, and file-based routing. While SvelteKit is powerful for complex applications, providing integrated solutions for SEO, fast loading times, and more, we decided against using it for our project. The primary reason is its complexity and the breadth of features it offers, such as SSR and SSG, which are unnecessary for our app's simpler requirements. Our focus is on leveraging Svelte's core capabilities for a lightweight and straightforward SPA, avoiding the overhead that comes with SvelteKit's extensive tooling and features.

Instead, in our app, we are going to use a simple `svelte-router` library to handle routing together with bare Svelte.

Let's now set up our project.

Setting up the project

We can create our Svelte project using the template that Vite provides. To do that, we can execute the following CLI command. It will ask us for a couple of configurations to set up the project further:

```
$ npm create vite
? Project name: › chat_frontend
? Select a framework:  Svelte
? Select a variant: TypeScript
```

Now, we can go to the newly created folder, add a routing dependency, and install the libraries by using the following commands:

```
$ cd chat_frontend
$ npm add axios@1.6.7 svelte-routing@2.11.0

$ npm install
```

We are going to use `axios` for sending API requests to our backend and `svelte-routing` for showing different pages based on the URL.

This setup has created a lot of files, so let's discuss the files that we see and tailor some of them to our needs.

Exploring the project structure

There are quite a lot of files that we got from the template setup in the previous section, so let's explore them.

Vite configuration file

The first file we will look into is the Vite configuration file. The Vite config is the place where you define settings and plugins that tailor how Vite builds and serves your application. Let's go through the important parts of this file:

vite.config.js

```
import { defineConfig } from 'vite'
```

The `defineConfig` function is used to create a configuration object for Vite.

The following line imports the Svelte plugin for Vite, enabling Vite to handle Svelte files:

```
import { svelte } from '@sveltejs/vite-plugin-svelte'
```

The plugin is responsible for compiling Svelte components and integrating them into the Vite build process.

This line exports the configuration object for Vite, using the `defineConfig` function:

```
export default defineConfig({
  plugins: [svelte()],
})
```

The object defines the settings and plugins that Vite will use. This setup has everything we need to develop the frontend part of our app. Now, let's explore the TypeScript configuration.

TypeScript configuration file

This file is used to configure TypeScript in the application, so let's look at it:

tsconfig.json

```
{
  "extends": @tsconfig/svelte/tsconfig.json,
```

By extending `@tsconfig/svelte/tsconfig.json`, we inherit a set of compiler options optimized for Svelte, ensuring that TypeScript plays nicely with Svelte's syntax and file structure.

This line specifies the `target` JavaScript version for the output code:

```
  "compilerOptions": {
    "target": "ESNext",
```

`"ESNext"` targets the latest supported **ECMAScript (ES)** features, allowing you to use the newest JavaScript features.

The following line ensures that class fields are defined using JavaScript's `defineProperty` function, aligning with the latest JavaScript standards for class field semantics:

```
"useDefineForClassFields": true,
```

The next code line determines the module system used in the project:

```
"module": "ESNext",
```

`ESNext` means the latest module syntax is used (such as `import` and `export`), which is standard in modern JavaScript development.

The next line allows importing JSON files directly into your TypeScript files, treating them as modules:

```
"resolveJsonModule": true,
```

This is useful for configuration files or other JSON data your application might need.

This enables JavaScript files to be included in your TypeScript project, allowing a mix of JavaScript and TypeScript files:

```
"allowJs": true,
```

The following tells TypeScript to type-check JavaScript files, offering the benefits of TypeScript's type system in regular JavaScript files:

```
"checkJs": true,
```

The next line ensures each file can be safely transpiled independently without relying on type information from other files:

```
"isolatedModules": true
```

This is important for certain build optimizations and is recommended when using Babel or when targeting ES modules.

The next lines include the source map files in the output build. Source maps help you to restore the precompiled step of your code. This is useful for debugging:

```
"sourceMap": true
```

The following line specifies which files are included in the TypeScript compilation context:

```
},
"include": ["src/**/*.ts", "src/**/*.js", "src/**/*.svelte"],
```

`include` includes TypeScript files, JavaScript files, and `*.svelte` Svelte files within the `src` directory, indicating a project structure where the source code is centralized in the `src` folder.

Our project uses project references to split the TypeScript configuration between different parts of the application. `tsconfig.node.json` is used to configure the local node environment to build our application, while `tsconfig.json` is used for browser support:

```
  "references": [{ "path": "./tsconfig.node.json" }]
}
```

The necessity for two distinct TypeScript configurations arises from the project's deployment across two separate execution environments:

1. The application code, located within the `src` folder, is designed to operate within a web browser environment. This setting leverages browser-specific APIs and functionalities.

2. The Vite configuration, along with its associated code, executes within Node.js on your local machine. This environment differs markedly from the browser, with its own set of APIs and limitations.

Let's now look into `tsconfig.node.json`.

The following config with `composite` enables project compilation in a way that allows TypeScript to more efficiently manage projects that are split into multiple build outputs:

tsconfig.node.json

```
{
  "compilerOptions": {

    "composite": true,
```

Then, the `skipLibCheck` option tells TypeScript to skip type checking of declaration files (`.d.ts` files from `node_modules`):

```
    "skipLibCheck": true,
```

This can speed up the compilation process by ignoring the types of third-party libraries.

Now, we will specify that we are going to run it with our Vite bundler:

```
    "module": "ESNext",
    "moduleResolution": "bundler"
```

The following code lines indicate that the config is specifically designed for type-checking and compiling the Vite configuration file:

```
    },
    "include": ["vite.config.ts"]
}
```

Let's turn our attention to the Svelte configuration.

Svelte configuration file

This configuration specifies the settings for Svelte. Let's take a look at it:

Svelte.config.js

```
import { vitePreprocess } from '@sveltejs/vite-plugin-svelte'
```

`vitePreprocess` is designed to work with Vite, facilitating the preprocessing of Svelte files.

Preprocessing can involve tasks such as transpiling TypeScript, handling SCSS or Less, and other file transformations before they're handed off to the Svelte compiler.

The following setup indicates that the Svelte compiler should use the preprocessing steps defined by `vitePreprocess` during the build process:

```
export default {
  preprocess: vitePreprocess(),
}
```

Let's now look at our `package.json` file.

Configuring package.json

This file serves as the main entry point for the general configurations in our applications; let's go through it:

package.json

```
{
    "name": "chat_frontend",
```

The next line prevents accidental publication of our project as a package to npm:

```
    "private": true,
```

The next script indicates that ES module syntax is used:

```
"version": "0.0.0",
"type": "module",
```

The following defines a script that starts the Vite development server:

```
"scripts": {
  "dev": "vite",
```

`vite build` builds the project for production using Vite:

```
"build": "vite build",
```

Now, we will write a config that starts a local web server that serves the built solution from `./dist`:

```
"preview": "vite preview",
```

The following code line runs `svelte-check` for linting and type-checking the Svelte components:

```
"check": "svelte-check --tsconfig ./tsconfig.json"
```

The next lines include development tools and dependencies such as Vite, the Vite plugin for Svelte (`@sveltejs/vite-plugin-svelte`), TypeScript, and others necessary for development but not required in production:

```
},
"devDependencies": {
  "@sveltejs/vite-plugin-svelte": "^3.0.2",
  "@tsconfig/svelte": "^5.0.2",
  "svelte-check": "^3.6.7",
  "tslib": "^2.6.2",
  "typescript": "^5.2.2",
  "vite": "^5.2.0"
},
```

The following section includes the libraries we need in production. We also moved the Svelte library here as we need it conceptually in production:

```
"dependencies": {
  "axios": "^1.6.7",
  "svelte": "^4.2.12",
  "svelte-routing": "^2.11.0"
}
}
```

Even though this config doesn't do much as all the libraries we use are going to be merged into single JavaScript files during the build, it is still useful to segregate which libraries we expect to be in the runtime for clarity.

Another file we can see is `package-lock.json`, which is the locked resolution of the dependencies from `package.json`.

We can now look at `index.html`, a static file that will be preserved in our build, which serves as our entry point TypeScript file.

HTML entry point

The following code block sets up the document structure, including meta tags for character encoding and responsive design, links a favicon, and defines the title of the page:

index.html

```
<!doctype html>
<html lang="en">
  <head>
    <meta charset="UTF-8" />
    <link rel="icon" type="image/svg+xml" href="/vite.svg" />
    <meta name="viewport" content="width=device-width, initial-
scale=1.0" />
    <title>Vite + Svelte + TS</title>
  </head>
  <body>
    <div id="app"></div>
    <script type="module" src="/src/main.ts"></script>
  </body>
</html>
```

The body contains a single `div` tag with an `id` property of `app`, which acts as the mounting point for the Svelte application. The application's entry point, `main.ts`, is linked as a module script, allowing the Svelte app to be initialized and rendered within the `app div` tag. This setup ensures that the application is correctly loaded and displayed in the browser.

We can also see the `public` folder in the root of our project tree, which we are going to use to store things that we can directly access from our HTML file, such as fonts, favicons, or tracking libraries.

Here is what the files in the `public` folder are responsible for:

- `src/vite-env.d.ts` is used to tell our TypeScript compiler to allow using environment variables from `import.meta.env`.

- `src/assets` is the folder that we are going to use to place our images and icons.

- `src/lib` is the place where we are going to put our Svelte components. We can remove the only component we see in this folder as we will write our own components here.

- `src/app.css` defines project-wide styles that should apply to all of our HTML components.

Now, let's explore the main entry point of our application.

Main file

The `main.ts` file is the entry point of our application. Let's go through this file.

The first line imports the CSS styles for the application. It ensures that the global styles defined in `app.css` are applied to the application:

main.ts

```
import './app.css'
```

We will import the main App component from `App.svelte`:

```
import App from './App.svelte'
```

The App component is the root component that wraps our entire application.

Here, we create and export a new instance of the App component:

```
const app = new App({
  target: document.getElementById('app')!,
})
export default app
```

The component is mounted to a DOM element identified by `id='app'` that we defined in our `index.html` file.

The last file that we need to cover is `src/App.svelte`. We are going to replace the content of the file so that it simply returns a header title for us.

Application title

We will replace the content of our main component with a very simple header:

src/App.svelte

```
<main>
  <h1>Chat</h1>
</main>
```

Now, if we execute the following command, we will see that our frontend server has started:

```
$ npm run dev
```

We will see a console output like this:

```
> chat_frontend@0.0.0 dev
> vite
  VITE v5.2.8  ready in 447 ms
  →  Local:   http://localhost:5173/
  →  Network: use --host to expose
  →  press h + enter to show help
```

If we visit the localhost server presented to us, we will see a big **Chat** title in the middle of the screen.

Now, you can see HMR in action by following these steps:

1. Add some additional HTML tags (such as `<h2>ai-based chat</h2>`) in `src/App.svelte`.

2. Save the `src/App.svelte` file.

3. You will see that the HTML tag is immediately visible in the browser.

With this, we have covered our frontend setup.

Summary

In this chapter, we have covered how to set up a frontend SPA project with Svelte, TypeScript, and Vite. We also learned what all the files in the configuration do, and now we are able to create a comfortable working environment with HMR and a fast-building tool.

In the next chapter, we will begin working on our frontend application functionality.

14

Svelte Chat Application Development

We've set up our Svelte development environment in the previous chapter, and now it is time to implement the logic of our frontend. We will cover topics such as routing in Svelte, developing the state management of our application, making API calls, and writing the overall logic of our chat. By the end of this chapter, you will be well equipped with tools to write your own frontend applications in Svelte.

We will cover the following topics:

- Writing routes for our application
- Handling authentication logic
- Developing chat logic
- Introducing styling

Technical requirements

To implement our chat functionality, we will need to install the following libraries:

```
$ npm install axios@1.6.7 jwt-decode@4.0.0 svelte-routing@2.11.0
```

We will use `axios` to make API calls, `jwt-decode` to get the payload from a JWT token, and `svelte-routing` to define the routing of our application.

All the code examples we are going to discuss in this chapter are available at https://github.com/PacktPublishing/Full-Stack-Web-Development-with-TypeScript-5/tree/main/Chapter14.

Writing routes for our application

Routing is a crucial aspect of web development, enabling navigation between different parts of an application without the need to reload the web page. It's essential for creating a seamless user experience and organizing the content structure within single-page applications. In Svelte, routing is not included out of the box. This means we need to integrate a third-party library to manage the routes.

For simplicity and ease of use, especially for beginners, Svelte routing is a practical choice. Svelte routing is straightforward to use and integrates smoothly with Svelte projects.

Let's use it to define the routes of our application that we will need. In the following code, we will define our routes for registration and signing in, and the routes we will use for our chats:

src/App.svelte

```
<script>
  import { Route, Router } from "svelte-routing";
  import Register from "./routes/Register.svelte";
  import Login from "./routes/Login.svelte";
  import Chat from "./routes/Chat.svelte";
</script>

<Router>
  <Route path="/register" component={Register} />
  <Route path="/login" component={Login} />
  <Route >
    <Chat chatId={null} />
  </Route>
  <Route path="/:id" let:params>
    <Chat chatId={params.id} />
  </Route>
</Router>
```

The `<Router>` component is the root component for handling routing. It wraps all `<Route>` components, which define different paths in the application. We need to wrap all our components with a router if we want to make them accessible by a specific URL.

Each `<Route>` component has a `path` prop that specifies the URL path for that route. When the URL matches the path of any `<Route>` component, the component specified in the component prop of the instance of `<Route>` is rendered.

The <Chat> component is rendered in two different routes:

- In the first case, it's rendered with a chatId prop of null when the URL is "/".

- In the second case, it's rendered with a chatId prop that comes from the URL parameters when the URL matches the /:id pattern. The let:params attribute is used to bind the URL parameters to a local variable parameter.

We need to specify two different routes for the chat depending on the presence of id because the library, unfortunately, doesn't provide the functionality to provide a default argument.

With this setup, we will be able to access all of our routes. Now, it's time to implement the components you see in the routes, and we will start with the authentication.

Handling authentication logic

To authenticate our users, we will need to do quite a few things:

1. First, we will need to define the logic for authentication and how we are going to store the authentication data in our application once a user logs in

2. Then, we need to write components for registration and logging in

3. And finally, we need to also redirect the user to the login page if they access a page that needs authentication and they are not logged in yet

Let's start with defining the logic for our authentication using a store in Svelte.

Defining the authentication store

The authentication store is going to be responsible for storing and providing our authentication token as well as providing user data from the JWT token payload. Let's write such a store:

src/stores/auth.ts

```
import { writable } from "svelte/store";
import axios from "axios";
import { jwtDecode } from "jwt-decode";

interface TokenPayload {
  name: string;
}
```

The preceding code block is the TypeScript structure of the data payload of our JWT token.

The next lines set a token as the default `Authorization` header for all `axios` requests so that the token is attached to every request we make with `axios` after we call the `setAxiosAuth` function:

```
function setAxiosAuth(token: string) {
    axios.defaults.headers.common["Authorization"] = `Bearer ${token}`;
}
```

The `createAuthStore` function is used to create the authentication store:

```
function createAuthStore() {
```

The `createAuthStore` function first checks whether there is an authentication token stored in the local storage. If there is, it sets this token as the default `Authorization` header for all `axios` requests, which serves as the holder of the authentication information for our server. Storing the token in the local storage makes sure that when we reload the page, our user is not going to lose the authentication capabilities.

The `createAuthStore` function then creates a `writable` Svelte store with the initial value set to the token from local storage:

```
const token = localStorage.getItem("authToken");
if (token) {
    setAxiosAuth(token);
}
const { subscribe, set } = writable<string | null>(token);
```

The `writable` function returns an object with `subscribe`, `set`, and `update` methods. These methods are used to subscribe to changes in the store, set the value of the store, and update the value of the store, respectively.

Then, the `createAuthStore` function returns an object with the `subscribe` method from the `writable` store and custom `set`, `remove`, and `getPayload` methods:

```
return {
    subscribe,
    set: (value: string) => {
        localStorage.setItem("authToken", value);
        set(value);
        if (value) {
            setAxiosAuth(value);
        } else {
```

```
        delete axios.defaults.headers.common["Authorization"];
      }
    },
```

When we export the `subscribe` method, we make an object to be a store in Svelte concepts and enable it to work with the Svelte reactivity.

The `set` method is used to set the value of the authentication token. It stores the token in local storage, sets the value of the Svelte store, and sets the token as the default `Authorization` header for all `axios` requests.

The `remove` method is used to remove the authentication token:

```
    remove: () => {
      localStorage.removeItem("authToken");
      set(null);
      delete axios.defaults.headers.common["Authorization"];
    },
```

The `remove` method removes the token from local storage, sets the value of the Svelte store to `null`, and removes the `Authorization` header from the `axios` defaults.

The `getPayload` method is used to get the payload of the JWT token. It decodes the token and returns the payload:

```
    getPayload: () => {
      const token = localStorage.getItem("authToken");
      if (token) {
        const decoded: TokenPayload = jwtDecode(token);
        return decoded;
      }
      return null;
    },
  };
}
```

Finally, the `authToken` store is created by calling the `createAuthStore` function:

```
export const authToken = createAuthStore();
```

This `authToken` store can then be used in other parts of our application to manage the authentication token.

Implementing Login component

We can write a Login component that will make it possible for our users to log in with their credentials.

Before we start working on a component, let's define a constant variable that will serve as a prefix for our backend API:

src/constants.ts

```
export const API_HOST = import.meta.env.VITE_API_HOST;
```

Here, we retrieve a variable from the environment called `VITE_API_HOST` and set it as a constant that we use when we make an API call to the application.

Vite will automatically pick up all the environments from the `.env` file that have a `VITE` prefix. Let's also define this environment file:

.env

```
VITE_API_HOST=http://localhost:3000
```

Let's proceed with the implementation of our Login component.

Providing a form for credential input

In this component, we will provide a form for email and password input, and then try to log the user in using the provided credentials.

Here are the main things we need to accomplish in the component:

- When users provide an email and password, try to log them in
- Validate that the email and password are not empty
- Gracefully handle the authentication errors coming back from the backend
- Save the user authentication token and redirect to our home page once we have successfully logged in

We will also import the styles for every component from our `styles` folder, and we will explore the contents of these files in the *Introducing styling* section of this chapter.

src/routes/Login.svelte

```
<script lang="ts">
  import { onMount } from "svelte";
  import { navigate } from "svelte-routing";
```

```
import axios from "axios";
import "../styles/auth.css";
```

The component has three reactive variables – `email`, `password`, and `errorMessage`:

```
import { authToken } from "../stores/auth";
import {API_HOST} from "../constants";

let email = "";
let password = "";
let errorMessage = "";
```

The `email` and `password` variables are bound to the corresponding input fields in the form, and their values change as the user types into these fields.

The `formValid` reactive statement checks that the `email` and `password` fields are not empty:

```
$: formValid = email.length > 0 && password.length > 0;
```

The `formValid` reactive statement is used to enable or disable the **Sign in** button. It gets reevaluated once the `email` and `password` values change.

The `onMount` function runs after the component is first rendered:

```
onMount(() => {
  if ($authToken) {
    navigate("/");
  }
});
```

The `onMount` function checks whether the user is already authenticated by checking the `authToken` store. If the user is authenticated, it navigates to the home page so that the user doesn't need to log in again when they are already authenticated.

The `login` function is an asynchronous function that is called when the user submits the form:

```
async function login() {
  try {
    const response = await axios.post(`${API_HOST}/api/v1/auth/
login/`, {
      email,
      password,
    });
    authToken.set(response.data?.token);
    navigate("/");
  } catch (error) {
```

```
      const defaultError = "An unexpected error occurred"
      if (axios.isAxiosError(error) && error.response) {
        const errorSlug = error?.response?.data?.error
        switch(errorSlug) {
          case "INVALID_CREDENTIALS":
            errorMessage = "Invalid email or password"
            break;
```

The login function sends a POST request to the /api/v1/auth/login/ endpoint with email and password as the request body. If the request is successful, it sets the authentication token in the authToken store and navigates to the home page. If the request fails, it sets the errorMessage variable based on the error received.

Here, we also check for a specific error of invalid credentials to show a more understandable error to the user.

Checking invalid credentials

The on:submit|preventDefault={login} directive is used to call the login function when the form is submitted. The preventDefault modifier is used to prevent the form from being submitted in the traditional way, which would cause the page to reload:

```
        default:
          errorMessage = defaultError
      }
    } else {
      errorMessage = defaultError
    }
  }
}

</script>

<div class="auth-container">
  <form on:submit|preventDefault={login} class="auth-form">
```

The errorMessage variable is used to display an error message to the user when the login attempt fails.

The error message is displayed in the form only if it's not an empty string:

```
    <div class="form-header">
      <h2>Login</h2>
    </div>
    {#if errorMessage}
```

```
    <div class="error">{errorMessage}</div>
  {/if}
```

Next, we use `bind` here as a two-way binding to our variables, so that when we type a new input, our `email` variable updates, and if we update `email`, our input updates. We do the same for the password field:

```
    <div class="input-group">
      <input type="email" placeholder="Email" bind:value={email}
required />
    </div>
    <div class="input-group">
      <input
        type="password"
        placeholder="Password"
        bind:value={password}
        required
      />
    </div>
    <div class="action-group">
```

If our form is not valid, we disable the button for the user; otherwise, this click triggers the form to submit:

```
    <button type="submit" class="auth-btn" disabled={!formValid}
      >Sign in</button
    >
  </div>
```

At the end of the login component, we also have a redirect link to the registration page if the user doesn't have an account yet:

```
    <div class="switch-auth">
      Don't have an account? <a href="/register">Register here</a>.
    </div>
  </form>
</div>
```

The registration process is the component that we are going to write next.

Implementing Register component

Our Register component will be somewhat similar in logic to our Login component, but it will have an additional input field, name, and we will call another endpoint to register a user.

We need to accomplish the following situation:

- When users provide their email, name, and password, try to register them with it
- Validate that the email, name, and password are not empty
- Gracefully handle the authentication errors coming back from the backend
- Redirect the users to the login page after the successful registration

Here is the code for the Register component:

src/routes/Register.svelte

```ts
<script lang="ts">
  import { onMount } from "svelte";
  import { navigate } from "svelte-routing";
  import axios from "axios";
  import "../styles/auth.css";
  import { authToken } from "../stores/auth";
  import {API_HOST} from "../constants";
```

Next, we declare the four reactive variables we are going to use in our form:

```ts
let name = "";
let email = "";
let password = "";
let errorMessage = "";
```

We follow a similar logic to onMount in the Login component and redirect the user to the home page if the user is already authenticated:

```ts
onMount(() => {
  if ($authToken) {
    navigate("/");
  }
});
```

Next, we try to register the user with their name, email, and password, and if it is successful, we redirect them to the login page. If we meet an error, we show it on the screen:

```
async function register() {
  try {
    await axios.post(`${API_HOST}/api/v1/auth/register/`, {
      name,
      email,
      password,
    });
    navigate("/login");
```

We handle a specific error that occurs when the user with the same email already exists; otherwise, we show a default error message:

```
  } catch (error) {
    const defaultError = "An unexpected error occurred"
    if (axios.isAxiosError(error) && error.response) {
      const errorSlug = error?.response?.data?.error
      switch(errorSlug) {
        case "ERROR_USER_ALREADY_EXIST":
          errorMessage = "User already exists, try logging in instead"
          break;
        default:
          errorMessage = defaultError
      }
    } else {
      errorMessage = defaultError
    }
  }
}
```

The `formValid` variable indicates that a user can submit a form:

```
  $: formValid = email.length > 0 && password.length > 0 && name.length > 0;
```

On submission of the form, we call the `register` function and prevent the page from reloading:

```
</script>

<div class="auth-container">
  <form on:submit|preventDefault={register} class="auth-form">
```

Here, we display the error message if it is present:

```
<div class="form-header">
  <h2>Create Account</h2>
</div>
{#if errorMessage}
  <div class="error">{errorMessage}</div>
{/if}
```

The next code lines show our input variables (name, email, and password), which are bound to the respective reactive variables:

```
<div class="input-group">
  <input type="text" placeholder="Name" bind:value={name} required
/>
</div>
<div class="input-group">
  <input type="email" placeholder="Email" bind:value={email}
required />
</div>
<div class="input-group">
  <input
    type="password"
    placeholder="Password"
    bind:value={password}
    required
  />
</div>
```

At the end, we have a redirect URL to the login page in case the user is already registered:

```
<div class="action-group">
  <button type="submit" class="auth-btn" disabled={!formValid}
    >Sign Up</button
  >
</div>
<div class="switch-auth">
  Already have an account? <a href="/login">Sign in here.</a>
</div>
</form>
</div>
```

Now, both our Login and Register components are defined and we can move to defining our chat-related components.

Developing chat logic

We will implement a few functionalities and components to support our chat logic so that, when our user is authenticated, we will allow them to do the following actions:

- Log out

- Create a new chat

- See all the created chats

- Select a created chat

- Load all messages in a chat

- Send messages inside a chat

We will begin with a parent component.

Creating a parent component

Let's start with a general component called Chat, which we will use when a user is authenticated; this is going to hold the other chat components that will accomplish the functionality mentioned previously:

src/routes/Chat.svelte

```ts
<script lang="ts">
    import {onMount} from "svelte";
    import {navigate} from "svelte-routing";
    import {authToken} from "../stores/auth";
    import ChatListSideBar from "../components/ChatListSideBar.
svelte";
    import ChatDetails from "../components/ChatDetails.svelte";
    import Header from "../components/Header.svelte";
    import '../styles/chat.css'
    export let chatId: string | null;
```

The chatId variable is declared as an exported variable, which means it can be passed as a prop from a parent component. In our case, it is passed by the routes. chatId is used to display the details of a specific chat.

In the onMount hook, we check whether the user is authenticated by checking the authToken store:

```
onMount(() => {
    if (!$authToken) {
        navigate("/register");
```

```
        }
    });
```

If the user is not authenticated, we navigate the user to the Register page.

The component's markup is divided into several sections, as shown here:

```
</script>

<div>
    <Header></Header>
    <div class="container">
        <div class="chat-list-container">
            <ChatListSideBar chatId={chatId}/>

        </div>
        <div class="chat-container">
            {#if chatId}
                <ChatDetails chatId={chatId}/>
            {/if}
        </div>
    </div>
</div>
```

The `Header` component is displayed at the top; this will display the user's name and give a button to log out. The `ChatListSideBar` component is displayed on the left side of the page and receives `chatId` as a prop. This will show all the chat and will provide a button to create a new one. The `ChatDetails` component is displayed on the right side of the page and also receives `chatId` as a prop. The `ChatDetails` component is only rendered if `chatId` is not `null`.

Let's now look at the implementation of the `Header` component.

Implementing the Header component

The `Header` component will retrieve the username from the JWT token payload and will remove the token from the store when the user logs out and redirect the user to the login page.

The `Header` component is the first component we define in the `components` folder. We have a separation for the structure clarity. All the components that are shown from the routes directly are defined in the `routes` folder, while all the components that are not rendered by the router are defined in the `components` folder:

src/components/Header.svelte

```ts
<script lang="ts">
    import { navigate } from 'svelte-routing';
    import { authToken } from '../stores/auth';
    import '../styles/header.css';
    const name = authToken.getPayload()?.name || "User"
```

We retrieve the name from the payload if it is not `null` or use `"User"` as the default name for display.

On logging out, we remove the saved JWT token and redirect the user to the login page:

```ts
    function logout() {
        authToken.remove();
        navigate('/login');
    }
</script>
```

In the next code lines, we define a simple structure to show the user's name and a button to log out:

```html
<div class="header">
    <div class="user-name">{name}</div>
    <button class="logout-button" on:click={logout}>
        Log out
    </button>
</div>
```

Now, let's focus on the component that is going to show the list of chats and also provide a button to create a new chat.

Creating the Chat component

The Chat component is going to retrieve all the chats from the server and also open a modal window where a user can create a new chat.

Retrieving chats from the server

In the following component, we will retrieve all of the existing chats and show them in a list on the left side of our screen.

src/components/ChatListSideBar.svelte

```ts
<script lang="ts">
  import { onMount } from "svelte";
  import axios from "axios";

  import { navigate } from "svelte-routing";
  import CreateChatPopup from "./CreateChatPopup.svelte";
  import {API_HOST} from "../constants";
  import '../styles/chatList.css'
  let chats: { id: string; name: string }[] = [];
```

The `chats` variable is an array that stores the list of chats fetched from the server.

The `errorMessage` variable is used to display an error message to the user when the `fetch` operation fails:

```ts
  let errorMessage: string | null = null;
```

The `chatId` variable is an exported variable. This means it can be passed as a prop from a parent component, which, for this component, means that the chat is selected:

```ts
  export let chatId: string | null;
```

The `getData` function is an asynchronous function that fetches the list of chats from the server:

```ts
  async function getData() {
    try {
      const response = await axios.get(`${API_HOST}/api/v1/chat/`);
      chats = response.data.data;
    } catch (error) {
      console.error("Error fetching chats:", error);
      errorMessage = "Failed to fetch chats. Please try again later.";
    }
  }
```

If the `fetch` operation is successful, the `getData` function updates the `chats` variable with the fetched data. If the `fetch` operation fails, the `getData` function updates the `errorMessage` variable with an error message.

When the `Chat` component first loads, we call the `getData` function to retrieve all the chats:

```ts
  onMount(async () => {
    await getData();
  });
```

The `isCreatingNewChat` variable is a Boolean that determines whether the `CreateChatPopup` component should be displayed:

```
let isCreatingNewChat = false;
```

The `selectChat` function is used to navigate to the chat with the given `chatId`:

```
function selectChat(chatId: string) {
  navigate(`/${chatId}`);
}
```

The preceding code lines will change our URL and provide our `Chat` component with the `chatId` variable, which will, in turn, pass `chatId` to all the children that use it.

The `createNewChat` function sets the `isCreatingNewChat` variable to `true`, which causes the `CreateChatPopup` component to be displayed:

```
function createNewChat() {
  isCreatingNewChat = true;
}
```

The `onCreate` function is called when a new chat is created from `CreateChatPopup`. It navigates to the new chat and fetches the updated list of chats:

```
async function onCreate(newChatId: string) {
  onClose();
  navigate(`/${newChatId}`);
  await getData()
}
```

The `onClose` function sets the `isCreatingNewChat` variable to `false`, which causes the `CreateChatPopup` component to be hidden:

```
function onClose() {
  isCreatingNewChat=false;
}
</script>
```

The following code structure allows the chat list sidebar to display a list of chats, navigate to a chat when it's clicked, display an error message when the `fetch` operation fails, and display a popup for creating a new chat:

```
<div class="chat-list-container">
  {#if errorMessage}
    <div class="error">{errorMessage}</div>
  {/if}
  {#if isCreatingNewChat}
    <CreateChatPopup onCreate={onCreate} onClose={onClose} />
  {/if}
  {#if chats.length === 0}
    <div class="no-chats">No chats available. Create a new one!</div>
  {/if}
  <div class="chat-list">
    {#each chats as chat (chat.id)}
      <div class="class-list-item" class:selected={chat.id === chatId}
on:click={() => selectChat(chat.id)}>
```

When the iterated chat ID matches the currently active chat obtained by the `chatId` prop, we add the `selected` class to highlight visually which chat is currently in use.

Let's show the chat's name and add a button for creating a chat:

```
        {chat.name}
      </div>
    {/each}
  </div>
  <button on:click={() => createNewChat()}>New Chat</button>
</div>
```

We can now look at the Popup component that creates a new chat by the typed-in name.

Creating a new chat

`CreateChatPopup` is used for creating a new chat by name; when we open the popup, the user is allowed to type a name for the chat, and then we try to create the chat. Once it is created, we close the popup:

src/components/CreateChatPopup.svelte

```
<script lang="ts">
  import axios from 'axios';
  import {API_HOST} from "../constants";
  import '../styles/chatPopup.css';
```

`onCreate` and `onClose` are provided as props from the parent, and we call them when we create a new chat or initiate the closure of the popup:

```
export let onCreate: (newChatId: string) => void;
export let onClose: () => void;
```

Doing this is an example of how we can easily share data and events between two coupled components.

We use the `chatName` and `errorMessage` variables to store the chat name in the input and the error that we can potentially get from the server:

```
let chatName = '';
let errorMessage: string | null = null;
```

The `createChat` function tries to create a chat, and if it is successful, it calls to its parent that a new chat is created by calling `onCreate` and providing the freshly created chat's ID:

```
async function createChat() {
  try {
    const response = await axios.post(`${API_HOST}/api/v1/chat/`, {
name: chatName });
    onCreate(response.data.data.id);
  } catch (error) {
    console.error('Error creating chat:', error);
    errorMessage = "Failed to create chat. Please try again later.";
  }
}
</script>
```

The component's markup is `div` with a class of `popup`, which we are going to use for styling:

```
<div class="popup">
  <div class="close-button" on:click={onClose}>X</div>
  {#if errorMessage}
    <div class="error">{errorMessage}</div>
  {/if}
  <input type="text" bind:value={chatName} placeholder="Enter chat
name" />
  <button disabled={!chatName.length} on:click={createChat}>Create</
button>
</div>
```

Inside the popup div, we have a button we use to close the popup that calls the onClose function when clicked, an error message that is displayed if there is an input field for the chat name, and a Create button that calls the createChat function when clicked. The Create button is disabled if chatName is an empty string.

The last component that we need to build is the component of a specific chat that will load the messages for a chat, allow us to type a new message, and show the response from our AI.

Showing chat details

ChatDetails is used to show messages in the chat. It loads all the messages from a specific chat and also allows us to send a new chat message in our chat:

src/components/ChatDetails.svelte

```ts
<script lang="ts">
  import { onMount } from "svelte";
  import axios from "axios";
  import {API_HOST} from "../constants";
  import '../styles/chatDetails.css'
  export let chatId: string;
  let messages: { message: string, createdAt: number }[] = [];
  let newMessage = "";
  let errorMessage: string | null = null;
  let isLoading = false;
```

The preceding code block lists these variables:

- The chatId variable is a prop of the selected chat.

- The messages variable is an array that stores the list of messages fetched from the server.

- The newMessage variable is bound to textarea in the form, and its value changes as the user types into this field.

- The errorMessage variable is used to display an error message to the user when the fetch or send operation fails.

- The isLoading variable is a Boolean that determines whether the "Send" button should be disabled. When we are waiting for the response from our server, which can take a while as we are waiting for GPT to generate a response, we will not allow sending new messages.

The `loadMessages` function is an asynchronous function that fetches the list of messages from the server:

```
onMount(async () => {
  await loadMessages();
});

async function loadMessages() {
  try {
    const response = await axios.get(
      `${API_HOST}/api/v1/chat/${chatId}/message/`
    );
    messages = response.data.data;
  } catch (error) {
    errorMessage = "Failed to get chat details. Please try again
later.";
    console.error("Error fetching messages:", error);
  }
}
```

If the `fetch` operation is successful, the `loadMessages` function updates the `messages` variable with the fetched data. If the `fetch` operation fails, it updates the `errorMessage` variable with an error message. We call it when the component first loads.

The `sendMessage` function is an asynchronous function that sends a new message to the server:

```
async function sendMessage() {
  isLoading = true;
  try {
    const response = await axios.post(
      `${API_HOST}/api/v1/chat/${chatId}/message/`,
      { message: newMessage }
    );
    messages = [...messages, {message:newMessage, createdAt: Date.
now()}, response.data.data];
    newMessage = "";
  } catch (error) {
    errorMessage = "Failed to send message. Please try again
later.";
    console.error("Error sending message:", error);
  }
  isLoading = false;
}
```

If the `send` operation is successful, the `sendMessage` function updates the `messages` variable with the new message the user typed and the response message from the server and then clears the `newMessage` variable. If the `send` operation fails, it updates the `errorMessage` variable with an error message.

The reactive statement with `$:` runs the `loadMessages` function whenever the `chatId` variable changes so that we get messages for a new chat when we choose a new chat:

```
$: {
  if (chatId) {
    loadMessages();
  }
}
</script>
```

The component's markup displays an error message if there is one, a list of messages with their creation times, `textarea` for typing a new message, and a `Send` button that calls the `sendMessage` function when clicked. The `Send` button is disabled if the `isLoading` variable is `true`:

```
<div class="chat-details-wrapper">
  {#if errorMessage}
    <div class="error">{errorMessage}</div>
  {/if}
  <ul>
    {#each messages as message (message.id)}
      <li>
        {message.message}
        <span>{new Date(message.createdAt).toLocaleTimeString()}</
span>
      </li>
    {/each}
  </ul>
  <textarea bind:value={newMessage} placeholder="Type a message"></
textarea>
  <button on:click={sendMessage} disabled={isLoading}>
    {#if isLoading}
      Sending...
    {:else}
      Send
    {/if}
  </button>
</div>
```

These are all the components that we need for our simple chat application to work.

Now, we can start our frontend server and observe the functionality of our application with the following command:

```
$ npm run dev
```

Among other lines, you will also see that our frontend server has started running on a certain port, something similar to this:

```
→   Local:    http://localhost:5173/
```

We need to add this URL to our backend-allowed CORS, as this is the URL from which we are going to access our backend. Add the provided frontend URL to the .env.dev file in the backend folder as a new environment variable.

```
CORS_ORIGIN=http://localhost:5173
```

Then, don't forget to start the backend server with the following command:

```
$ bun --env-file .env.dev dev
```

At this point, you can start using our frontend application. I encourage you to first try to register and log in, then create a new chat, and try to send a couple of messages to our server and have a small conversation with our LLM API.

Well, this is all amazing, but it looks pretty basic and bland, so let's spice it up a little bit with our styles.

Introducing styling

We've already included our styles in the right places in the components, so what is left for us to do is to actually implement the styling. We will start with app.css, which we import into our main file, which means that it's going to be applied to all of our components in the application. We will put there the styles that we want to be shared across all the components.

Writing application-wide styling

The following code lines mean that the children of the form will be laid out in a column, one below the other:

src/app.css

```css
form {
  display: flex;
  flex-direction: column;
}
```

The following lines give the input fields a certain amount of space around them, padding inside them, a border, and rounded corners:

```css
input {
  margin-bottom: 10px;
  padding: 8px;
  border: 1px solid #ccc;
  border-radius: 4px;
}
```

The button style sets several properties to style the buttons:

```css
button {
  padding: 10px;
  background-color: blue;
  color: white;
  border: none;
  border-radius: 4px;
  cursor: pointer;
}
```

```css
button:hover {
  background-color: #0876fa;
}
```

```css
button:disabled {
  background-color: gray;
}
```

We add padding and background-color, the color of the text, and remove the border. The border-radius property is set to create a slight rounding. The cursor property is set to pointer, which changes the cursor to a pointer when hovering over the button. There are also hover and disabled states defined for the button, which slightly change the color.

The .error class makes our errors' text red and bold:

```
.error {
  color: red;
  font-weight: bold;
  margin-bottom: 10px;
}
```

Now, let's look at the styles we use for our Login and Register components.

Styling for our Login and Register components

Here, we are going to discuss the styles we use to make our authentication forms look smoother:

src/styles/auth.css

```
.auth-container {
    display: grid;
    place-items: center;
    min-height: 100vh;
    background-color: #f4f7fa;
}
```

This style centers all content within .auth-container both horizontally and vertically using CSS Grid. The container extends to the full height of the viewport (100vh), ensuring that the authentication form is centered on the page. The background color is a light grayish blue, providing a neutral, calming background.

The .auth-form class applies to the authentication form, setting a maximum width of 400px and full width as default to ensure it works responsively on different devices:

```
.auth-form {
    width: 100%;
    max-width: 400px;
    padding: 20px;
    box-shadow: 0 4px 6px rgba(0, 0, 0, 0.1);
    border-radius: 8px;
    background: white;
}
```

Padding is added for internal spacing, a subtle box-shadow for depth, rounded borders for aesthetics, and a white background to stand out against the lighter container background.

The following lines target <h2> elements inside .form-header, providing a bottom margin for spacing and centering the text enhancing readability and focus:

```
.form-header h2 {
    margin-bottom: 20px;
    text-align: center;
}
```

.input-group is used for grouping form elements, providing a consistent bottom margin that separates each input group for clarity:

```
.input-group {
    margin-bottom: 20px;
}
```

Inputs within .input-group take the full width of their parent, have padding for comfort in typing, rounded borders for a modern look, a light gray border for definition, and box-sizing: border-box to include padding and border in the element's total width and height:

```
.input-group input {
    width: 100%;
    padding: 15px;
    border-radius: 4px;
    border: 1px solid #ccc;
    box-sizing: border-box;
}
```

.action-group centers text inside of itself:

```
.action-group {
    text-align: center;
}
```

.auth-btn is styled for form submission buttons; these are given full width, substantial padding, and no border with rounded edges:

```
.auth-btn {
    width: 100%;
    padding: 10px 15px;
    border: none;
    border-radius: 4px;
    background-color: #5C6BC0;
    color: white;
    font-size: 16px;
```

```
    font-weight: bold;
    cursor: pointer;
    transition: background-color 0.3s;
}
```

The background is a deep blue, the text color is white for contrast, and the font is bold and slightly larger for easy reading. The cursor changes to a pointer to indicate it's clickable, with a background color transition for a responsive hover effect.

Both the `hover` and `disabled` states share a darker blue shade, providing immediate visual feedback on interaction or state change:

```
.auth-btn:hover,
.auth-btn:disabled {
    background-color: #3F51B5;
}
```

When the button is disabled, the cursor reverts to the default to reflect the non-interactive state.

```
.auth-btn:disabled {
    cursor: default;
}
```

This `.switch-auth` class centers the text inside and the content above it:

```
.switch-auth {
    margin-top: 20px;
    text-align: center;
}
```

Within the `.switch-auth` section, spans are styled in a medium gray, with a right margin for spacing:

```
.switch-auth span {
    color: #666;
    margin-right: 5px;
}
```

The links in `.switch-auth` are bold, in the same blue color as buttons for consistency, with no underline to keep a clean look:

```
.switch-auth a {
    color: #5C6BC0;
    text-decoration: none;
```

```
    font-weight: bold;
    transition: color 0.3s;
}
```

A color transition is also applied for a subtle hover effect.

On hover, the link color darkens and an underline appears, enhancing user interaction cues by making the link visually responsive:

```
.switch-auth a:hover {
    color: #3F51B5;
    text-decoration: underline;
}
```

We will now add styles for the header component.

Styling our header component

The `.header` class is applied to the topmost part of a layout, commonly used as a navigation or status bar:

src/styles/header.css

```
.header {
    display: flex;
    height: 5vh;
    justify-content: flex-end;
    align-items: center;
    padding: 10px 20px;
    background-color: #f4f7fa;
    border-bottom: 1px solid #ccc;
}
```

This styling sets the header's height to 5% of the viewport height (5vh), ensuring that it takes minimal vertical space while still being noticeable. The styling uses Flexbox to align items, with `justify-content: flex-end;` placing child elements (such as user controls or logout buttons) toward the right-hand side of the header.

`align-items: center;` ensures that all child elements are vertically centered, making the header content appear vertically aligned. Padding is added on all sides (`10px` top and bottom, `20px` left and right) to give the content some breathing room. The background color is set to a light grayish blue, creating a calm and neutral tone, and a bottom border of light gray (`#ccc`) adds a subtle visual separation from the rest of the content.

Then, the `.user-name` class targets the element displaying the user's name. It is pushed to the left side of the flex container due to `margin-right: auto;`, which automatically assigns all remaining horizontal space to the right of this element. This ensures that the username is prominently displayed at the start of the header:

```css
.user-name {
    margin-right: auto;
    padding: 10px;
    font-size: 16px;
    color: #333;
}
```

Padding is uniformly set to `10px` for comfort, preventing any text from sticking too closely to the edges. The font size is moderately set to `16px` for clear readability, and the text color is dark gray (`#333`), offering good contrast against the lighter background of the header, making the username easy to spot.

Now, let's look at the styles we use for the chat container.

Styling the chat container

The style options for our chat container set up a container using the Flexbox layout, stretching it to cover 90% of the viewport height (`90vh`):

src/styles/chat.css

```css
.container {
    display: flex;
    height: 90vh;
}
```

The use of Flexbox ensures that its child elements (in this context, likely chat lists and chat windows) are aligned in a row and can be sized and spaced responsively.

The `.chat-list-container` class is designed for the part of the interface that holds the list of chats. It is given a fixed width of 20% of its parent container and does not grow or shrink (`flex: 0 0 20%`):

```css
.chat-list-container {
    flex: 0 0 20%;
    display: flex;
    flex-direction: column;
    justify-content: space-between;
}
```

This chat list container uses a vertical Flexbox layout (`flex-direction: column`), which arranges its children (such as individual chat list items) in a column. The `justify-content: space-between;` property distributes the child elements evenly with space between them, aligning the first item to the top and the last item to the bottom, making it visually organized and easy to navigate.

The `.chat-container` class applies to the area where chat messages are displayed. It is designed to take up the remaining space in `.container` not used by `.chat-list-container`:

```
.chat-container {
    flex: 1;
}
```

The `flex: 1;` property allows this container to grow and fill the space.

We can now talk about the styles we use for the chat list.

Styling for the chat list

The `.chat-list` class is applied to a list element that displays a list of chat entries:

src/styles/chatList.css

```
.chat-list {
    list-style: none;
    padding: 0;
    margin: 0;
    max-height: 95vh;
    overflow-y: auto;
}
```

This chat list style removes the default list styling (`list-style: none`) and sets both `padding` and `margin` to 0 to ensure that the list fills its container without unnecessary space. It limits the height of the chat list to 95% of the viewport height (`95vh`), allowing for a controlled scrolling area (`overflow-y: auto`), which activates vertical scrolling when the content exceeds the maximum height.

Each list item within `.chat-list` has a padding of `15px` for a spacious touch area, making each chat entry easier to interact with:

```
.chat-list-item {
    padding: 15px;
    border-bottom: 1px solid #ddd;
    cursor: pointer;
    transition: background-color 0.3s ease;
    border-radius: 3px;
```

```
    margin: 5px;
}
```

A subtle border (1px solid #ddd) at the bottom of each item helps visually separate one chat entry from another. The cursor is set to pointer to indicate that these items are clickable. A background color transition effect is added for smooth visual feedback when interacting with the list items. Rounded corners (border-radius: 3px) soften the visual design, and a small margin keeps items visually distinct from each other.

The .no-chats class is used for a message displayed when there are no chats available:

```
.no-chats {
    color: #999;
    text-align: center;
    margin-top: 20px;
    margin-bottom: 20px;
}
```

The .no-chats class styles the text color in a light gray (#999), which indicates a less important or inactive state. Text is centered, and vertical margins provide ample spacing from other content, focusing attention on the message in the absence of chat entries.

Buttons within .chat-list-container are styled to expand to the full width of their container, ensuring that they are easily clickable and maintain a consistent appearance:

```
.chat-list-container button {
    width: 100%;
}
```

Hovering over chat list items changes their background to a bright blue (#579ae8), with the text color switching to white, enhancing readability and user interaction by clearly highlighting the active selection:

```
.chat-list-item:hover {
    background-color: #579ae8;
    color: white;
}
```

The .selected class, applied to a chat list item that is currently active or opened, mirrors the hover style, maintaining a consistent visual cue that this item is selected. It has the same bright blue background and white text, ensuring it stands out against other list items:

```
.selected {
    background-color: #579ae8;
    color: white;
}
```

Let's look at the styling for our chat popup creation component.

Styling our chat popup creation component

The .popup class is used for modal dialogs, positioned in the center of the viewport using a combination of position: fixed, top: 50%, left: 50%, and transform: translate(-50%, -50%). This ensures that the popup is always centered, regardless of the viewport size:

src/styles/chatPopup.css

```css
.popup {
    position: fixed;
    top: 50%;
    left: 50%;
    transform: translate(-50%, -50%);
    width: 300px;
    padding: 20px;
    background-color: white;
    border-radius: 8px;
    box-shadow: 0px 0px 10px rgba(0, 0, 0, 0.1);
    z-index: 1000;
    display: flex;
    flex-direction: column;
    justify-content: center;
    align-items: center;
}
```

The fixed width and padding provide structure and internal space, while the white background and rounded borders create a clean, friendly appearance. A light box-shadow adds depth and prominence, and z-index ensures the popup remains above all other content. The Flexbox layout with column direction centers its child elements vertically and horizontally.

Text inputs within the popup are styled to take up 80% of the popup's width, with padding for comfortable typing and rounded corners for a modern look:

```css
.popup input[type="text"] {
    width: 80%;
    padding: 10px 5px;
    border-radius: 4px;
    margin-bottom: 10px;
}
```

A margin at the bottom separates the input from other elements, helping to maintain visual clarity and focus.

Buttons in the popup are slightly wider than the text inputs, filling 85% of the popup's width, which helps them stand out as primary interaction elements:

```
.popup button {
    width: 85%;
    padding: 10px;
}
```

The padding ensures they are easy to interact with, catering to usability.

A close button positioned absolutely within the popup allows it to float in the top-right corner, which makes it easily accessible for users to dismiss the popup:

```
.popup .close-button {
    position: absolute;
    top: 10px;
    right: 10px;
    cursor: pointer;
}
```

The cursor pointer on hover indicates that the element is interactive, enhancing the user experience by making it clear that clicking the button will perform an action, specifically, closing the popup.

Now, let's cover the last set of styles for our specific chat.

Styling for specific chats

The .chat-details-wrapper class is styled to organize elements within a chat interface vertically:

src/styles/chatDetails.css

```
.chat-details-wrapper {
    display: flex;
    flex-direction: column;
    justify-content: space-between;
    height: 100%;
    padding-bottom: 10px;
}
```

By using display: flex and flex-direction: column, the classs arranges its children in a vertical stack. justify-content: space-between ensures that the first child is aligned to the top and the last child is aligned to the bottom, with the maximum space between them. The wrapper takes the full height of its container (`height: 100%`) and has padding at the bottom for spacing away from any adjacent elements.

This style applies to `<ul>` elements, commonly used to list items such as chat messages:

```
ul {
    list-style: none;
    padding: 0;
    margin: 0;
    overflow-y: auto;
    flex-grow: 1;
}
```

`list-style: none` removes bullet points, and `padding` and `margin` set to 0 remove default spacing to maximize content area. `overflow-y: auto` enables vertical scrolling when the content exceeds the element's height. `flex-grow: 1` allows the instance of `<ul>` to expand and fill the space in flex containers, making it responsive to varying amounts of content.

Each list item (`<li>`) within a instance of `<ul>` has padding for spacing inside, a light gray border at the bottom (#ddd) to visually separate individual items, and vertical margins to space out list items slightly from each other, enhancing readability:

```
li {
    padding: 10px;
    border-bottom: 1px solid #ddd;
    margin: 5px 0;
}
```

`textarea` is styled to not resize, maintaining a consistent layout that includes padding for comfortable typing, a solid border, and slightly rounded corners for aesthetics:

```
textarea {
    resize: none;
    padding: 10px;
    border: 1px solid #ccc;
    border-radius: 4px;
    margin-bottom: 10px;
}
```

A margin at the bottom separates `textarea` from any subsequent content, such as a `send` button.

Now, our application looks much more decent!

With this, we have completed writing the main logic of our frontend application.

Summary

In this chapter, we have learned how to write real-world-like applications using Svelte. We have covered how to use routing, apply styling, handle authentication, use reactivity in real cases, use styling, and interact with the backend. This forms the foundation for our ability to write frontend applications with Svelte.

In the next chapter, we are going to talk about more advanced topics of Svelte such as linting, testing, and internationalization.

15

Advanced Svelte Techniques

In this last chapter of the book, we are going to cover the more advanced aspects of Svelte: formatting, linting, internationalization, accessibility, and testing. All of these topics are essential to writing a robust and user-friendly frontend application. By the end of this chapter, you will have a solid grasp of developing frontend applications in general.

We are going to cover the following topics:

- Configuring formatting and linting
- Exploring a11y
- Setting up i18n
- Introducing testing
- Further reading

Technical requirements

To introduce the functionality mentioned in the introduction, we will need to install a few additional libraries. First, we must install the libraries that are required for formatting and linting. These are `eslint` and `prettier`, and they should be familiar to you from *Chapter 3*, where we introduced linting and formatting for our backend application:

```
$ npm install --save-dev @typescript-eslint/eslint-plugin@^7.7.0 @
typescript-eslint/parser@^7.7.0 eslint@^8.57.0 eslint-config-
prettier@^9.1.0 eslint-plugin-prettier@^5.1.3 eslint-plugin-
svelte@^2.37.0 prettier@^3.2.3 prettier-plugin-svelte@^3.2.3
```

Let's talk about each library:

- `@typescript-eslint/eslint-plugin`: This plugin integrates TypeScript language features with ESLint, allowing for TypeScript code to be linted using ESLint rules.

- `@typescript-eslint/parser`: An ESLint parser that enables ESLint to lint TypeScript code, parsing TypeScript files into a format that ESLint can understand.

- `eslint`: The ESLint library we use for linting that runs static checks against our code to detect issues such as unused variables.

- `eslint-config-prettier`: A config that disables rules that might conflict with Prettier, ensuring that ESLint and Prettier do not give conflicting formatting advice.

- `eslint-plugin-prettier`: This plugin integrates Prettier into ESLint by running it as a rule within ESLint. It identifies and reports formatting discrepancies as individual ESLint issues.

- `eslint-plugin-svelte`: Adds support for Svelte components in ESLint, allowing for linting of Svelte-specific syntax and patterns.

- `prettier`: The Prettier library for formatting our code according to the standards we choose.

- `prettier-plugin-svelte`: This plugin formats Svelte components using Prettier, ensuring that Svelte code is formatted according to Prettier's standards.

We will also need to install some libraries to test our application:

```
$ npm install --save-dev vitest@^1.5.0 jsdom@^24.0.0 @testing-library/
jest-dom@^6.4.2 @testing-library/svelte@^5.0.1
```

Let's take a closer look at each library:

- `vitest`: A fast, Vite-based testing framework that's designed to provide a Jest-like experience with improved performance, which is well-integrated with Vite

- `jsdom`: A JavaScript implementation of web standards such as DOM and HTML, allowing Node.js to simulate a browser environment for testing

- `@testing-library/jest-dom`: This library extends `vitest` with custom matchers to simplify asserting conditions on DOM nodes in tests

- `@testing-library/svelte`: This library provides utilities to functionally test Svelte components, encouraging best practices in testing

Finally, we will need to install a library for internationalization:

```
$ npm install svelte-i18n
```

We will use `svelte-i18n` to translate our application into different languages.

All the code examples we are going to discuss in this chapter are available at `https://github.com/PacktPublishing/Full-Stack-Web-Development-with-TypeScript-5/tree/main/Chapter15`.

Configuring formatting and linting

Just like we did for the backend, in our frontend, we need to ensure that our application follows the same standards across the whole project. We also want to try to catch potential issues with our code early. To simplify this process, we can automate our formatting and linting using ESLint and Prettier. To glue everything together, we will need to add additional plugins that integrate the libraries with Svelte since Svelte has its own syntax, and therefore requires additional plugins to handle its code.

First, we will configure ESLint.

ESLint config file

Let's start by discussing our main ESLint config file.

.eslintrc.cjs

We'll begin with `env`, which specifies the environments the script is expected to run in:

```
module.exports = {
  env: {
    browser: true,
    es2021: true,
  },
```

Here, it's set for browser environments and acknowledges the use of ECMAScript 2021 features.

The `extends` key is used to inherit configurations from a set of predefined `recommended` configurations:

```
  extends: [
    "plugin:svelte/recommended",
    "eslint:recommended",
    "plugin:@typescript-eslint/recommended",
    "prettier",
    "plugin:prettier/recommended"
  ],
```

This setup extends ESLint configurations for Svelte, TypeScript, and Prettier. This helps enforce best practices and style consistency.

The `parserOptions` config sets the parser's behavior:

```
parserOptions: {
```

The `ecmaVersion` config indicates that the code uses ECMAScript 2021 features:

```
ecmaVersion: 12,
```

The `sourceType` config signals that the code uses ES modules:

```
sourceType: "module",
```

The `project` config specifies the TypeScript configuration file path:

```
project: "./tsconfig.json",
```

The `extraFileExtensions` config allows the parser to handle files with the `.svelte` extension, which are Svelte components:

```
extraFileExtensions: [".svelte"],
```

The following section allows different ESLint settings for specific file types:

```
},
overrides: [
```

The following line shows that the override is specific to Svelte component files – that is, `*.svelte`:

```
{
  files: ["*.svelte"],
```

The following lines configure ESLint to correctly process and parse Svelte files:

```
processor: "svelte/svelte",
parser: "svelte-eslint-parser",
```

The `parserOptions` config parses scripts within Svelte files that use TypeScript:

```
parserOptions: {
  parser: "@typescript-eslint/parser",
```

The `rules` config defines specific rules for ESLint to enforce or ignore:

```
      },
    },
  ],
  rules: {
```

The following line turns off the rule that prevents you from declaring functions or variables within nested blocks, which is a common pattern in JavaScript:

```
"no-inner-declarations": "off",
```

The following line ensures that any formatting inconsistencies found by Prettier are reported as errors:

```
      "prettier/prettier": "error",
    },
};
```

Now, let's define the rules for Prettier that ESLint will use when it tries to lint the files.

Prettier file

The following file specifies configurations for our Prettier `plugins` key in it:

.prettierrc

```
{
    "plugins": ["prettier-plugin-svelte"],
```

Here, `prettier-plugin-svelte` is included, which adds support for the Svelte syntax for Prettier. This allows Prettier to correctly format Svelte component files.

The following lines ensure that the Svelte-specific syntax is handled correctly during formatting:

```
    "overrides": [{ "files": "*.svelte", "options": { "parser": "svelte"
} }]
}
```

Now, let's add an additional command to the `scripts` section of our `package.json` file to handle linting. We will also add a command for testing that we will need later in the *Introducing testing* section:

package.json

```
{
    "name": "chat_frontend_3",
    "private": true,
    "version": "0.0.0",
    "type": "module",
    "scripts": {
        "dev": "vite",
        "build": "vite build",
        "preview": "vite preview",
```

```
    "check": "svelte-check --tsconfig ./tsconfig.json",
    "lint": "eslint src --fix",
```

The `lint` command runs `eslint` in our `src` folder and fixes any issues it can fix automatically, such as styling.

The following line starts running our tests using `vitest`:

```
    "test": "vitest"
  },
...
```

Now, we can run the following command:

```
$ npm run lint
```

We will see that a lot of files have been adjusted so that they have the correct styling and that there are a couple of errors in our project:

```
/chat_frontend_3/src/components/ChatListSideBar.svelte
  61:7   error  A11y: visible, non-interactive elements with an
on:click event must be accompanied by a keyboard event handler.
Consider whether an interactive element such as <button type="button">
or <a> might be more appropriate. See https://svelte.dev/docs/
accessibility-warnings#a11y-click-events-have-key-events for more
details.(a11y-click-events-have-key-events)  svelte/valid-compile
  61:7   error  A11y: Non-interactive element <li> should not be
assigned mouse or keyboard event listeners.(a11y-no-noninteractive-
element-interactions)
/chat_frontend_3/src/components/CreateChatPopup.svelte
  25:3   error  A11y: visible, non-interactive elements with an
on:click event must be accompanied by a keyboard event handler.
Consider whether an interactive element such as <button type="button">
or <a> might be more appropriate. See https://svelte.dev/docs/
accessibility-warnings#a11y-click-events-have-key-events for more
details.(a11y-click-events-have-key-events)  svelte/valid-compile
  25:3   error  A11y: <div> with click handler must have an ARIA
role(a11y-no-static-element-interactions)
✘  4 problems (4 errors, 0 warnings)
```

All these errors are about accessibility. Let's look at this in more detail.

Exploring a11y

Accessibility is essential for creating digital products that can be used by everyone, including those with disabilities. It's not only a matter of ethical practice and inclusivity but also a legal requirement in many parts of the world. Accessible websites and applications can reach a broader audience, provide a better user experience, and often result in cleaner and more maintainable code.

The importance of accessibility

Accessibility is important for several reasons:

- **Inclusivity**: Accessibility ensures that all users, regardless of their physical or cognitive abilities, can access and interact with content. This includes people with visual impairments, hearing difficulties, motor limitations, and cognitive disorders.

- **Legal compliance**: Many countries have regulations requiring digital content to be accessible. For example, the **Americans with Disabilities Act** (**ADA**) in the US and the European Accessibility Act in the EU set standards that websites must meet.

- **SEO benefits**: Accessible sites tend to be better optimized for search engines. Features that improve accessibility, such as image alt text and proper heading structures, can also enhance SEO performance.

- **Improved usability**: Accessible design improves the overall user experience and usability of a website for all users, not just those with disabilities.

Svelte facilitates some of these, so let's talk about what it helps us with.

Accessibility rules and best practices in Svelte

Svelte facilitates various best practices through its reactive and declarative syntax. Here are a few ways Svelte helps with accessibility:

- **a11y warnings**: Svelte's compiler will warn us about missing alt text on images and non-interactive clickable elements

- **Role and ARIA properties**: Svelte allows us to easily integrate **Accessible Rich Internet Applications** (**ARIA**) roles and properties to enhance the accessibility of complex web components

- **Animation and motion reduction**: Svelte provides tools to respect the `prefers-reduced-motion` CSS media feature, reducing animations for users who have motion sensitivity

Now, let's see how we can improve accessibility in our application.

Fixing accessibility issues in our project

After executing `npm run lint`, we'll see that we have issues in the `CreateChatPopup` component, mainly with our closing button. Let's take a closer look at these issues:

- **Keyboard inaccessibility**: The element cannot be focused or activated using the keyboard. Users who rely on keyboard navigation (such as those using screen readers or those unable to use a mouse) would be unable to interact with this button.

- **Semantic incorrectness**: The `div` element does not semantically convey that it is a button, which can confuse assistive technologies that rely on the correct roles to provide context to users.

Let's improve our close button element in `src/components/CreateChatPopup.svelte`. We need to add additional attributes, `role` and `tabindex`, to enhance a11y. To do so, we will edit our `div` element and add the `role` and `tabindex` attributes. This will fix our a11y issues:

```
<div class="close-button" on:click={onClose}>X</div>
```

Replace the preceding line with the following code block:

```
<div
    class="close-button"
    on:click={onClose}
    on:keydown|preventDefault={(e) => e.key === "Enter" && onClose()}
    role="button"
    tabindex="0"
  >
    X
  </div>
```

To solve our a11y issues, we added the following:

- `tabindex="0"`: This attribute makes the element focusable using the keyboard, specifically allowing it to be focused by the keyboard's *Tab* key navigation. The `"0"` value means that the element can be reached in the order it appears in the HTML source, fitting seamlessly into the sequence of table elements on the page.

- `on:keydown|preventDefault={(e) => e.key === "Enter" && onClose()}`: This adds a keyboard event listener that triggers the `onClose` function when the *Enter* key is pressed. The `|preventDefault` modifier is used to prevent any default behavior that might be triggered by pressing the *Enter* key, ensuring the button's functionality is isolated to closing only. This allows users who do not use a mouse to activate the button with the *Enter* key, mimicking how native buttons behave.

- `role="button"`: By assigning the role of `"button"` to the `div` element, this code explicitly tells assistive technologies that the `div` element is meant to be used as a button. This helps screen readers and other assistive tools understand the purpose of the element, thus providing a correct description to users who rely on this information for navigation and interaction.

Moving forward, to fix an issue that's similar to the close button, we will need to go to the `src/components/ChatListSideBar.svelte` file and change our chat lists so that the accessibility attributes are present. To do this, we will replace the following code lines:

```
<ul class="chat-list">
    {#each chats as chat (chat.id)}
      <li
        class="chat-list-item"
        class:selected={chat.id === chatId}
```

```
        on:click={() => selectChat(chat.id)}
    >
        {chat.name}
    </li>
    {/each}
</ul>
```

Replace the preceding code with the following:

```
<div class="chat-list">
    {#each chats as chat (chat.id)}
        <div
            class="chat-list-item"
            class:selected={chat.id === chatId}
            on:click={() => selectChat(chat.id)}
            on:keydown|preventDefault={(event) =>
                event.key === "Enter" && selectChat(chat.id)}
            tabindex="0"
            role="button"
        >
            {chat.name}
        </div>
    {/each}
</div>
```

Here, we replaced our list element with `div` elements and added the `chat-list-item` class to them so that we can handle the tab index and role. These elements should not be assigned to `li` as *non-interactive elements cannot have an interactive role button*.

With this, we have fixed our linting and accessibility issues, and we are ready to talk about internationalization.

Setting up i18n

Internationalization, often abbreviated as **i18n**, refers to the practice of designing and building software in such a way that it can be easily adapted to different languages and regional settings without the need to alter the underlying code structure. Internationalization is a critical aspect of global software development because it allows for the localization of content, formats, and functionality to meet the specific cultural and linguistic needs of different target audiences. It typically involves using local time, language, order of letters, and even culture-specific designs and colors.

Localization, or **l10n**, is a subset of internationalization. It mostly refers to translations of the content. This is what we are going to focus on in this chapter as the most important part of i18n. In the context of Svelte, translations can be efficiently managed using the `svelte-i18n` library. This library provides a straightforward and powerful way to integrate dynamic localization capabilities into Svelte applications. So, let's learn how to integrate it.

First, we must define our translation files. As an example, we will translate a string into both English and Ukrainian.

Here is the `logout` key in English:

src/locales/en.json

```
{
    "logout": "Logout"
}
```

Here is the `logout` key in Ukrainian:

src/locales/ua.json

```
{
    "logout": "Вийти з системи"
}
```

Here, we defined a similar `logout` key in two files, so our library will choose the correct specific translation based on the language used. Now, we can initialize our translation.

We need to set `i18n` up before we can use it in our application. Once we've done that, we will need to define it before the other elements of our app so that when we import it, it's ready to use.

We can import the required functions from `svelte-i18n`:

src/i18n.ts

```
import { init, getLocaleFromNavigator, addMessages } from "svelte-
i18n";
```

Next, we must import the files with the specific translations and register them in the system by their full locale names:

```
import ua from "./locales/ua.json";
import en from "./locales/en.json";

addMessages("en-US", en);
addMessages("uk-UA", ua);
```

In the following code block, we're initializing our library, providing the default locale that we will use if we don't have a translation for the user's locale, and also trying to get the current user locale from the browser settings:

```
init({
  fallbackLocale: "en-US",
  initialLocale: getLocaleFromNavigator(),
});
```

Now, let's import our translation into the main file:

src/main.ts

```
import "./i18n";
...
```

Here, we imported our translation functionality into the file to make sure it is evaluated first in our system.

Now, we are ready to use i18n. Let's try to implement a language switcher that will change the language of our app. Since we've only defined the translations of the word logout, let's focus on handling i18n and show the correct language for the logout button. To do that, we need to edit our Header component, which contains our logout button:

src/component/Header.svelte

```
<script lang="ts">
  import { navigate } from "svelte-routing";
  import { authToken } from "../stores/auth";
  import "../styles/header.css";
  import { _, locale } from "svelte-i18n";
```

Here, we import the necessary functions. First, _ is used to show a key in the correct translation. After this, we have locale. This is a Svelte store value that we can bind to an input field and update the locale that's used in the system:

```
  const name = authToken.getPayload()?.name || "User";
  function logout() {
    authToken.remove();
    navigate("/login");
  }
</script>

<div class="header">
  <div class="user-name">{name}</div>
```

Here, we will provide a `select` element that binds to the `locale` value. When we change the input, it will be reflected in the setup language in our system:

```
<div class="locale">
  <select bind:value={$locale}>
```

We can indicate that the `logout` key is in the correct locale based on what the user's current locale is:

```
    <option value="en-US">English</option>
    <option value="uk-UA">Ukrainian</option>
  </select>
</div>
  <button class="logout-button" on:click={logout}>{$_("logout")}</button>
</div>
```

If you open the application on the home screen and try to change the language, you will see how our translation of the `logout` button changes on the fly. We have more content in the application to translate, so you can follow a similar process to translate other strings in the project.

Now, we are ready to cover the last aspect of this chapter: testing.

Introducing testing

Two popular types of testing can occur on the frontend:

- **Component/unit testing**: Unit testing is easier to write in the frontend setup, helps us ensure higher overall code coverage, and allows us to test in isolation
- **End-to-end testing**: End-to-end testing ensures that our code works and that the integration between pieces is smooth

We will focus on unit testing in this section to perform the higher code coverage, but I encourage you to also read up about setups for end-to-end testing using Svelte and different testing techniques here: `https://svelte.dev/docs/faq`.

We will start by configuring Vite for testing.

Configuring the Vite file

First, we need to expand the configuration for our Vite file so that it supports testing. We can do this by using the following code:

vite.config.ts

```
import { defineConfig } from "vite";
import { svelte } from "@sveltejs/vite-plugin-svelte";
import "vitest/config";
```

Now, we must import the `vitest/config` library to enable a new key in the config called `test`:

```
// https://vitejs.dev/config/
export default defineConfig({
  plugins: [svelte()],
  test: {
    globals: true,
```

This tells Vitest to automatically inject global variables into the test files. This makes common testing functions such as `describe`, `it`, and `expect` available in every test file.

`environment` specifies the environment in which the tests will run:

```
    environment: "jsdom",
```

`jsdom` is a common choice for projects that need a simulated DOM environment as it allows us to test DOM interactions without a browser. This simulates a web browser's environment so that DOM-related functionality can be tested.

`setupFiles` lists files that will be executed before the tests run:

```
    setupFiles: ["./src/setupTests.ts"]
});
```

In `setupFiles`, you can place the global setup code that needs to be run before any tests are executed. This could include setting up mocks, configuring global variables, or other preparatory tasks that are necessary for the tests to run correctly.

Next, we'll discuss what's in our setup file.

Exploring our setup file

Here, we must import dom, which allows us to execute functions such as `toBeInTheDocument()` on our `expect` statements that are specific to various the dom elements:

src/setupTests.ts

```
import "@testing-library/jest-dom/vitest";
```

Next, we must import `vitest` itself:

```
import { beforeAll, vi } from "vitest";
```

The following code is a hack that fixes an unfortunate bug between Svelte and Vite that exists at the time of writing – without these lines, Vite won't be able to execute the life cycle hooks of the Svelte component:

```
// @ts-expect-error - svelte/internal is a module, wrong error
import * as svelteinternal from "svelte/internal";
beforeAll(() => {
  vi.mock("svelte", () => svelteinternal);
});
```

The preceding code also contains `ts-expect-error`, which we use to suppress an incorrect TypeScript error on the line. This error indicates that `svelte/internal` is not a module, even though we know it is.

Now that you understand the beauty of the JavaScript World Library's integrations, we can start writing a test for one of our components.

Writing a test for our component

In this section, we will write a couple of tests for `ChatListSideBar.svelte`. Here, we will provide some chat data to it, throw an error, press a button to create a new chat, and test that whatever is rendered is what we expect to be rendered.

To do this, we must import the Svelte-specific function that will help us imitate an event, render our element in a fake DOM tree, and wait until some text finally renders in the DOM.

src/components/ChatListSideBar.test.ts

We'll put the test next to the original component since this is a common convention in the frontend world. However, it is not required:

```
import { beforeEach, describe, expect, it, vi } from "vitest";
import { fireEvent, render, waitFor } from "@testing-library/svelte";
```

Next, we must import external dependencies that we will mock to prevent the API call and router navigation:

```
import { navigate as navigateOriginal } from "svelte-routing";
import axios from "axios";
```

The following code lines import the component we are going to test and mock dependencies for:

```
import ChatListSideBar from "./ChatListSideBar.svelte";
vi.mock("axios");
vi.mock("svelte-routing", () => ({
  navigate: vi.fn(),
}));
```

axios and svelte-routing are mocked in the following code block. Mocks replace these modules with test-specific functions, allowing us to gain control over their behavior without relying on external services:

```
const navigate = vi.mocked(navigateOriginal, true);
const axiosGet = vi.mocked(axios, true).get;
```

Next, we must create some dummy data for our chats. We can provide this as the mock API response when we retrieve chats:

```
describe("ChatListSideBar", () => {
  const mockChats = {
    data: {
      data: [
        { id: "chat1", name: "Chat 1" },
        { id: "chat2", name: "Chat 2" },
      ],
    },
  };
```

We must clear our mocks in every run. This can be seen in the following code snippet. This ensures that mocks are reset before each test runs so that each test starts with a fresh state:

```
beforeEach(() => {
  axiosGet.mockClear();
  navigate.mockClear();
});
```

The following test checks that an error message is displayed if chats fail to load due to an API error. This can be simulated by rejecting axios.get with an error:

```
it("displays an error message when chats fail to load", async () =>
{
  axiosGet.mockRejectedValue(new Error("Failed to fetch chats"));
```

We're mocking `axios` so that it doesn't imitate the API error.

`findByText` is used to check if some text is rendered in our tree:

```
const { findByText } = render(ChatListSideBar);
```

`waitFor` tries to call `findByText` again if it throws an error, but fails if it continues throwing errors beyond the timeout:

```
await waitFor(
```

Next, we must check that the element with the text of the failed request is present on the screen. The test will fail if the body doesn't succeed in `100` ms:

```
async () => {
  const errorMessage = await findByText(
    "Failed to fetch chats. Please try again later.",
  );
  expect(errorMessage).toBeInTheDocument();
},
{
  timeout: 100
},
);
});
```

Here, we provide an empty chat response from `axios` and ensure we show a message indicating that no chats are on the screen:

```
it("displays no chats message when there are no chats", async () =>
{
  axiosGet.mockResolvedValue({ data: { data: [] } });
  const { findByText } = render(ChatListSideBar);
  expect(
    await findByText("No chats available. Create a new one!"),
  ).toBeInTheDocument();
});
```

This is our happy path test. It checks that the chats are rendered correctly when we get them from the API.

The following code imitates the process of navigating to a specific chat:

```
it("displays chats when data is loaded", async () => {
  axiosGet.mockResolvedValue(mockChats);
```

```
      const { findByText } = render(ChatListSideBar);
      expect(await findByText("Chat 1")).toBeInTheDocument();
      expect(await findByText("Chat 2")).toBeInTheDocument();
  });
```

Now, let's imitate the process of clicking on a chat element to see whether it navigates to a specific URL:

```
  it("navigates to the chat when a chat item is clicked", async () =>
{
      axiosGet.mockResolvedValue(mockChats);
      const { findByText } = render(ChatListSideBar);
      const firstChatItem = await findByText("Chat 1");
      expect(firstChatItem).toBeInTheDocument();
      await fireEvent.click(firstChatItem);
```

The preceding code checks that our mock value was called with the correct URL.

Next, we must test that the nested component gets rendered when we want to create a new chat. This part has more of an integration test nature than a unit test nature but without the external dependencies:

```
      expect(navigate).toHaveBeenCalledWith("/chat1");
  });
```

We can imitate the process of clicking on the new chat button like so:

```
  it("shows create chat popup when new chat button is clicked", async
() => {
      axiosGet.mockResolvedValue({ data: { data: [] } });
      const { getByText, findByText } = render(ChatListSideBar);
      await fireEvent.click(getByText("New Chat"));
```

Finally, let's check that part of the popup is visible in our render tree:

```
      expect(await findByText("Create")).toBeInTheDocument();
  });
});
```

At this point, we have showcased how we can test different UI edge cases and how UI elements interact. With this, we are ready to conclude this section.

Further reading

There are a lot of other topics in the frontend and Svelte world that you can benefit from understanding. If you wish to learn more, I highly recommend exploring the following resources:

- **UI kits**: UI kits provide ready-made components with consistent styling and a lot of customization. They are useful for speeding up development, though you're limited in what you can achieve with them. I recommend checking out **Headless UI** (`https://svelte-headlessui.goss.io/docs/2.0`) and **SvelteUI** (`https://svelteui.dev/`).

- **Server-side rendering (SSR)**: SSR can be a very useful technique as it speeds up how quickly users see content and makes web pages more searchable by search engines. It does this by generating the HTML on the server before sending it to the user's browser. I highly recommend checking it out at SvelteKit (`https://kit.svelte.dev/`). It has built-in features such as server-side rendering, routing, and smooth client-side transitions.

- **Animations**: Animations can greatly enhance how good and smooth an application looks by providing dynamic styling. You can follow the official tutorial from Svelte to find out more: `https://learn.svelte.dev/tutorial/tweens`.

Summary

In this chapter, we covered more advanced aspects of Svelte development, such as linting, formatting, localization, accessibility, and testing. All of these aspects are required in all-rounded and production-like applications and at this point, we know how to handle them effectively.

With this, we have come to the end of this book. I hope it has been a good learning journey for you and that you have a much better grasp of how to develop an end-to-end application that includes frontend, backend, and external integrations by using cutting-edge technologies.

If you've enjoyed this book, I'd be extremely flattered to see a review on the Amazon page for this book.

I hope you have a fun and exciting coding journey – best of luck! ;)

Index

packtpub.com

Subscribe to our online digital library for full access to over 7,000 books and videos, as well as industry leading tools to help you plan your personal development and advance your career. For more information, please visit our website.

Why subscribe?

- Spend less time learning and more time coding with practical eBooks and Videos from over 4,000 industry professionals

- Improve your learning with Skill Plans built especially for you

- Get a free eBook or video every month

- Fully searchable for easy access to vital information

- Copy and paste, print, and bookmark content

Did you know that Packt offers eBook versions of every book published, with PDF and ePub files available? You can upgrade to the eBook version at packtpub.com and as a print book customer, you are entitled to a discount on the eBook copy. Get in touch with us at customercare@packtpub.com for more details.

At www.packtpub.com, you can also read a collection of free technical articles, sign up for a range of free newsletters, and receive exclusive discounts and offers on Packt books and eBooks.

Other Books You May Enjoy

If you enjoyed this book, you may be interested in these other books by Packt:

TypeScript 4 Design Patterns and Best Practices

Theofanis Despoudis

ISBN: 978-1-80056-342-1

- Understand the role of design patterns and their significance
- Explore all significant design patterns within the context of TypeScript
- Analyze, and develop classical design patterns in TypeScript
- Find out how design patterns differ from design concepts
- Understand how to put the principles of design patterns into practice
- Discover additional patterns that stem from functional and reactive programming

Learn React with TypeScript

Carl Rippon

ISBN: 978-1-80461-420-4

- Gain first-hand experience of TypeScript and its productivity features
- Understand how to transpile your TypeScript code into JavaScript for running in a browser
- Build a React frontend codebase with hooks
- Interact with REST and GraphQL web APIs
- Design and develop strongly typed reusable components
- Create automated component tests

Packt is searching for authors like you

If you're interested in becoming an author for Packt, please visit `authors.packtpub.com` and apply today. We have worked with thousands of developers and tech professionals, just like you, to help them share their insight with the global tech community. You can make a general application, apply for a specific hot topic that we are recruiting an author for, or submit your own idea.

Share Your Thoughts

Now you've finished *Full-Stack Web Development with TypeScript 5*, we'd love to hear your thoughts! Scan the QR code below to go straight to the Amazon review page for this book and share your feedback or leave a review on the site that you purchased it from.

https://packt.link/r/1835885594

Your review is important to us and the tech community and will help us make sure we're delivering excellent quality content.

Download a free PDF copy of this book

Thanks for purchasing this book!

Do you like to read on the go but are unable to carry your print books everywhere?

Is your eBook purchase not compatible with the device of your choice?

Don't worry, now with every Packt book you get a DRM-free PDF version of that book at no cost.

Read anywhere, any place, on any device. Search, copy, and paste code from your favorite technical books directly into your application.

The perks don't stop there, you can get exclusive access to discounts, newsletters, and great free content in your inbox daily

Follow these simple steps to get the benefits:

1. Scan the QR code or visit the link below

https://packt.link/free-ebook/9781835885581

2. Submit your proof of purchase
3. That's it! We'll send your free PDF and other benefits to your email directly

www.ingramcontent.com/pod-product-compliance
Lightning Source LLC
LaVergne TN
LVHW080114070326
832902LV00015B/2576